Fire Suppression/EMS
In the Shadow of Racial Disparity

Fire Suppression/EMS

In the Shadow of Racial Disparity

Theodore R. Coleman

Theodore R. Coleman
Fort Washington, MD

Photographs used by permission of the District of Columbia Fire and Emergency Medical Services Department.

Cover Design by TLH Designs, Chicago, IL, www.tlhdesigns.com

Layout and design by Kingdom Living Publishing, Fort Washington, MD, www.kingdomlivingbooks.com

Published by: Theodore R. Coleman
                 Fort Washington, MD 20744

ISBN: 978-0-615-91709-2

Published in the United States of America.

Dedication

To my wife Uvaghn, a dedicated partner for life. She bore our five children, and helped me in making our home a place for joy.

To my children, Andre', Theodore III, Michael, Sandra, and Yvette, thanks for being so respectful towards your mother and me, at all times.

To my father, Theodore R. Coleman, Sr. (deceased), and mother, Essie Mae New Coleman (deceased), thank you for all your love and caring for me in my young life.

To my brother, James G. Coleman, and sister, Hazel Coleman Miller (deceased), thank you for our lives together.

To my stepfather, Jordan R. Hairston, thank you for caring for our mother, Essie Mae Coleman-Hairston.

To my grandchildren, aunts, uncles, nieces, nephews, and cousins, may the love of our family tradition continue to abound in you always.

Acknowledgments

I would like to thank the following persons:

Therman King, the person who listened while I talked, read through thousands of documents given to him, and fashioned the information, as co-author, for publishing of this book.

Alice B. King, the typist of written material from her husband Therman.

Uvaghn Coleman, my wife, who would remind me of the details of many subject matters that I had to address in being a fireman.

My colleagues, too many to mention. The Fire Department would not have progressed as it did during my position at the helm, if it were not for people like you, you, and you.

Table of Contents

Chapter 1: The Early Years   13

Chapter 2: Employment at the U.S. Naval Gun Factory   24

Chapter 3: Appointment as a Firefighter D.C.   27
     Fire Department

Chapter 4: Mandatory Training at the Fire Academy   35

Chapter 5: Incidents of Discrimination and Racism   40
     in the D.C. Fire Department

Chapter 6: Black Firefighters Detail Assignment   46

Chapter 7: Promotion to the Rank of Sergeant   51

Chapter 8: Promotion to the Rank of Lieutenant   54

Chapter 9: Promotion to the Rank of Captain   59

Chapter 10: Assignment to the Rescue Squad   63

Chapter 11: Promotion to Battalion Fire Chief   65

Chapter 12: Promotion to Deputy Fire Chief,   67
     Assignment to Training Academy,
     and Acting Assistant Fire Chief of Services

Chapter 13: Assuming the Fire Chief Position   77

Chapter 14: From the Top Down Frankly Speaking   92

Chapter 15: The Fire Department Budget   97

Chapter 16: Success of Operating Programs   102

Chapter 17: The Ambulance Crisis   105

Chapter 18: The Snow Episode   118

Chapter 19: No Confidence Vote of Fire The Chief   123

Chapter 20: Protest to the Promotion Issue   127

Chapter 21: Standing on the Shoulders of Others   139

Chapter 22: Magazine Publication   145

Chapter 23: The Fireman's Parade   153

Chapter 24: The Missing Photo   159

Chapter 25: Retirement from Fire Department   189

Chapter 26: Reflections   226

Epilogue   232

Appendix A: A Concise History of the D.C.   234
         Fire Department

Appendix B: Seminole Indian Scouts   249

Appendix C: No Confidence Theory   253

Preface

I began to wonder sometime ago about the role a form of fate may have impacted in my life. It seemed that it was all I could think about, so I discussed my constant thoughts with my wife Uvaghn and later with other family members. They all agreed that I should write of my life's experience, especially about my 36 years in the Washington, District of Columbia, Fire Department. As I looked back in an effort to try to determine when and where memorable events took place, I realized there are no clearly cut beginnings to the inside story of my life. Introspection, however, has a significant affect in setting into motion a sequence of events whose outcome could never have been foreseen.

Hopefully, reflections of events in my life that I am about to describe will remind some readers that reflections of their early remembrance and experience in life may allow, if necessary, a refocusing and a renewed reminder of important situations in their lives. Hopefully, that will establish the when, where, and why's in determining a logical approach to understanding all circumstances, while believing in the saying, "The secret to contentment is the realization that life is a gift and not a right." Also, "No good deeds go unpunished."

My events will hopefully inspire anyone who reads of my experiences to realize that life in itself is without a doubt a physical and mental challenge that may affect each of us differently. However, one must continue in a state of grace until it is succeeded by a state of glory, better known as perseverance.

I began as a private rank in the District of Columbia Fire Department and retired as Fire Chief, having held the position for 6 years, longer than any other Fire Chief in the history of the department. I managed a $100 million dollar budget annually and 1,535 employees.

The D.C. Fire Department once lacked any and all meaningful forms of discourse by key responsible personnel. They seemed not to have any logical or reasoning abilities to communicate in difficult situations, in a formal and orderly communication, via speech or writing. This posturing demonstrated a wavering effect by persons in authority, as well as the rank and file, which was physically confrontational at times. There was a lack of clarity in the day-to-day duty mission of the Fire Department in Washington, D.C. It was as if the department was operating in the shadow of darkness.

During my position as Fire Chief, one of my primary objectives was to stress civil conduct behavior among the rank and file department personnel, in an effort to abate any form of misconduct. I would be remiss in not stating that other staff had tried doing the same in the past. The changes that I saw prior to my retirement were as distinguishable as night and day, in developing the department in being one as second to none.

Chapter 1

The Early Years

I was born in Danville, Va., to the parents of Theodore R. Coleman and Essie Mae (New) Coleman. My father was the son of Alexander and Mozell Coleman. They all lived in Ellington, S.C. Had it not been for the love, guidance, and affection given by my parents and grandparents, early life in development may have been difficult for me. I am profoundly grateful to all of them.

My early reflections are of my parents, grandparents, brother James, sister Hazel, and my developing years in youth until the age that I was drafted into the United States Army.

My father was the second of four siblings and was the most disciplined by his mother, rather severely at times because she did not spare the rod. His father, Alexander Coleman, worked for a wealthy white family named Evensberry. His wife Mozell and their four children lived on Mr. Evensberry's land. In doing so, he was building a new house for Mr. Evensberry and his family. My grandfather was proficient in many skills including brick masonry, cement finishing, carpentry, plumbing, and farming. He was also a Baptist Minister. He subsequently converted to the House of Prayer under the teaching of "Sweet Daddy Grace" of the Pentecostal faith. He spoke to Mr. Evensberry about the need for some land to build a house for his growing family. Mr. Evensberry told my grandfather that he knew of some land that was up for tax sale and that he would purchase it, and in turn, sell it to him. The site was 55-plus acres. My grandfather and his son Alex completed the house; then they began building the house for themselves. It was referred to as "the big house." The land surrounding the house was rich soil, ideal for farming.

My father was old enough and capable of doing farm work. He would work from "can to can't" (dawn to dusk), while his father and uncle continued building houses for other people, as a means of providing for the needs of their families.

When my father turned 18 years of age, he became increasingly disenchanted with farm work and decided to leave home. The train tracks ran approximately 300 yards of their land. One Sunday evening while the family was having supper, he slipped out of the house and hopped on the side of a Charleston & Western Carolina (C&WC) railroad car. His destination was wherever it would take him. When the train stopped the next day, he saw signs to indicate that he was in a town in the state of Virginia. He knew no one in the town, was quite hungry, and had very little change in his pocket. He began looking for work and a place to stay at the same time. He was certain that the town's people were wondering who he was; however, he went into a café belonging to a colored person and told an older employee his name, where he was from, and that his father was a Minister in Ellington, S.C. The man offered him some food and advised him to go back home. He returned home to Ellington, S.C., the same way he left, by being a hobo on the train without paying a fare. When he returned home, his family was glad to see him.

For the next three years, he and his horse (Charlie) plowed the fields where the crops were raised. During that time as well, he and his father started building a house for him in the anticipation of his getting married. Shortly after the house was finished, he married my mother, Essie Mae New.

My father and grandfather built a small building on the side of the family house that was initially used as a shoe repair shop. It was not as financially successfully as they would have liked, so my grandfather agreed for him to convert it to a clothing dry cleaners. It was the only dry cleaners in the little town for people of African descent to patronize. Subsequently, my father was successful in getting a job on the Charleston & Western Carolina railroad as a cook for the section crew workers. The income from this employment helped in further developing the dry cleaners. My father and his sister agreed that she would run the business during the weekdays, and he would help on Friday evenings and all day on Saturday. According to the processing evaluation, the dry cleaning method used was extremely hazardous. They used gasoline to clean the clothes in a tub, and then hung them on a line to dry in the outdoors on non rainy days. Then they were pressed to perfection with a steam press. The steam boiler that was used was developed and installed by my grandfather's self-creative ingenuity.

My father decided that he would work to accumulate enough money to invest in another business, such as a restaurant, since his self-styled cooking skills were very much in evidence by the remarks from the railroad crew members. A man named Mr. Reed, who was associated with the railroad, offered him a job making twice the salary that he was making, if he would go to work in his restaurant in Danville, Va. After working for the railroad for a time, he was ready to venture out. He and his sister agreed that she would maintain the cleaning establishment, while he and his wife went to Danville, Va., in pursuit of a restaurant. In the meantime, he decided to look up Mr. Reed, who he met while on the railroad, who had offered him a job. He found him and Mr. Reed kept his promise of giving him a job in his restaurant. A short time later, Mr. Reed opened a new restaurant in Durham, N.C., and asked him if he would work there. He agreed, and worked there for a considerable number of years. Eventually, with the help of Mr. Reed, he was able to get his own restaurant. He called it the "Coleman's Café. In addition, he acquired a barbershop, a poolroom, and a fleet of taxicabs. All of these businesses were located in Durham, N.C.

I was very young when the restaurant was operated. I had many playmates, because I had chewing gum all the time. When I was 12 or so, my father and mother divorced. Therefore, my father sold his business, and my sister, brother, and I returned to my father's home place in Ellington, S.C. His father died a few years later. I missed my grandfather and the fun I had with him playing around the house. I remembered the previous summer visits; it was always nice and warm, with butterflies on every flower plant. When it rained, the earth smelled so pure and clean. I attended public schools, and worked in my father's shoe repair shop that he had restored in place of the dry cleaners.

Of all other routines in the Coleman household, we would attend church regularly. One Sunday I asked my father if I could miss church. He looked at me curiously, and asked me why. Before I could give him a definitive answer, he said okay. I can only conclude that his thoughts were that I might have wanted some time to think something through. I really wanted to shine shoes for patrons of our shoe repair shop, so that I could have some extra spending money. It was a known fact that the shirts of colored men may not be without wrinkles, but you could bet your last dollar that their shoes would be shining.

In the winter months, the weather was chilly, but never too cold. As a heating source in the shop, we had a kerosene heater rather than a fireplace or potbelly stove. This particular Sunday morning, it was unusually cold and I knew the churchgoers would be coming in soon to get their shoes shined, so I lit the kerosene heater to warm the shop up. A few people came in to get their shoes shined, and continued on their way. After a while, things got quiet, so I decided to sit on the shoeshine stand and read the Sunday paper. I sat for a while and then noticed kerosene running from the heater. It caught fire. The heater had a one-gallon supply tank with an automatic shutoff valve at the bottom. The valve malfunctioned, causing excessive kerosene to reach the burners and overheat. The heater and floor became engulfed in flame. My first thought was to get the heater to the outside of the shop. I looked for a stick or broom handle, or something to use in getting it out of the door, but there was nothing in sight. Kerosene at that time was as explosive as some gasoline is today. With my last burst of energy, I managed to kick the heater toward the door, and the tank fell from the back of the heater, causing kerosene to splash all over the place. I then mustered the energy to kick the heater at the top, and it rolled out of the door and over on its side. I was determined to get the fire out; because a similar fire had taken place years before on the other side of the tracks and caused a number of houses to be burned up. Had I not gotten the heater outside, and then got the fire out on the inside, there would have possibly been a repeat of the destruction on this side of the tracks. The nearest fire department was 25 miles away in Augusta, Ga.

One other time I had a brush with fire that was more dramatic and very costly. It almost cost me my freedom. My friend and I were rabbit hunting, and the dogs ran a rabbit across the road into the tall grass. My dog would chase rabbits only if he could see them, so I decided to set the grass on fire so he could see them. I had seen grass burned before in clearing of fields, and I thought 400 feet between the road and the other side was far enough for the fire to remain on our side. Boy was I mistaken. The wind caused the grass to burn so furious that there was no way we could stop it. I had no idea how far the fire would burn, or who owned the land. I went home and told my father what happened. He said that it was the magistrates place. We got in the car and rode to the edge of the river. The fire took on a new meaning; it had continued to burn to the Savannah River, which separates the state of South Carolina from the state of Georgia.

My father looked scared that day. Both of us were afraid of the consequences surrounding this incident. We were notified by the town authority to be at the magistrates house at 10:00 a.m. the next day. Shortly after 10:00 a.m. that morning, I was sworn in. The prosecuting attorney asked me questions about the fire. After some 45 minutes, he said, "I knew your grandfather, and your father, they are both good people, and you have no blemishes on you, and let's keep it at that." He then hit the kitchen table with his gavel, and said, "Case adjourned; pay 10.00 dollars for the cost of the court. Case dismissed."

I thought about what he said at my court hearing that morning, "The next time anybody burns off this land, they're (go-in) to jail, (go-in) to the chain gang." He scolded me, and I was now repentant for being spared harsher punishment, thanks in large measure to my father's rapport with the landowner. Normally, that meant a friendly chat with him, or the Police Chief, and manager of the plant where bushel baskets were made. I never knew whether any town officials ever associated with the Klan. Whispers of accusation from my elders erased all reasonable doubt. If grandfather or father knew, they never made it my business. It was said some officials found few Negroes worthy of anything more than a day's pay.

Farming never thrilled me, but in Ellington, S.C., in the 1930's, cotton, corn, and cows were as bountiful as the flowing 150-foot high cypress trees sprouting along the red clay banks of the Savannah River. Only through some miracle would one escape a life of farming. I was excited with the saxophone. That may have been a way out of Ellington for me. If not for the movies with black entertainers and seeing Cab Calloway and Lena Horne entertaining through music and song, there was nothing to link Ellington to the world, or for the rest of the state for that matter. To see that not all blacks worked for the white man, at least that is how it looked gazing at the screen in the movie house, spawned hope.

In Ellington, the land stretched to the fiery tips of each morning's rising sun. Life never went beyond its borders. If it could not be grown on the sprawling acres, it likely was not worth acquiring. Everything was self-contained. Even the dreams of black folk stopped at the border.

There were no visible cross burnings or public rallies. The lack of Klan symbols made life less threatening. Suspected Klansmen were thought to be people who cared about us, because we plowed their land or raised their livestock, and their children.

Church attendance on Sunday was mandatory. Saturday night was the devil's day, and no decent Negro would dare patronize the speakeasies that were as plentiful as the cypress, or at least that was grandfather's edict. Though few outside the family heeded him on avoiding sin, grandfather was the community oracle. If a race problem surfaced, he would handle it. My father Theodore Sr., or 'Thado' or 'Fado,' to those finding his name difficult to master, nurtured a tolerance borne of his love of the land as an independent farmer and grandfather's admonishments to find the good in all men no matter how evil they may be. Maintaining unshakeable dignity was a Coleman trademark. I acquired these traits much later in my life, for it seemed, if my father and grandfather managed to command so much respect, it naturally transferred to me. I emulated practically everything they did.

A man said to my father one day, "Thado, look a-here, when that boy get out of school he won't have to serve in the war if he can get a deferral approval from the Military Draft Board, based on his services needed here on the farm." For white landowners, this was less of a humanitarian gesture, and more of recognition that their livelihood would be crippled by the loss of field hands. Scores of Ellington's black boys died old men struggling to repay the debt. Most often, surviving meant squelching all notions of independence, or ambition, leaving black men destined to be farmhands.

After graduating from high school, I got a job at the Lech Banana Case Company that was approximately 2 miles away from home. It employed many colored people. Bushel baskets were made at this business referred to as, "the plant." Machines were used to make the baskets. A 1,500-basket quota was considered our day's work. If you were proficient, you could complete your quota in six hours. Many of us would do it, and spend the remainder of our shift in the cafeteria near the plant. One day after making my 1,500 hundred bushel baskets, my friend and I began running the 2 miles back home, only to stop and rest at Mr. Turner's Café, before continuing on our way.

After dinner, my stepmother told me that I had some mail. I picked up the letter from my desk. It seemed that of a business type. It was that all right, a date and time for me to report for military duty, signed by the President of the United States. I was inducted into the Army at Fort Jackson, S.C. On our way there, I did not think we were going to make it by the way the driver drove the bus. He drove the bus up and down hills, around curves in such a reckless way I began to wonder if he had been selected to kill everyone on board in order to not have to waste time training us. After that most terrifying ride, I was ready to do anything they told us to do; however, we were able to engage in social and recreational activities in between our routine of processing. After a few days of such, we were transported to the Army Base at Fort Bragg, N.C., for more orientation, and then to Fort McClellan, Ala. This is where the shoe rubber hit the road. It was referred to as the worst hellhole of the Army. It was rainy, muddy, very cold at night, and then 100-degree temperature during the day. Everywhere we went while in training was up and down hills. Being a country boy was a blessing, for I was one tough 150-pound 18-year old. After basic training, we were put on a little ragged boat, and off to Japan we went. I became sea sick not too long after the boat got underway. I had never been sea sick in my life. My sickness was so serious; it caused me to wish for the worst kind of thing to happen to us. The 17-day trip on that smelly boat was a bit much to endure.

After my tour in Japan, I returned to the United States by boat and was assigned to an army depot, in San Diego, Calif., pending further assignment. Some soldiers were given a weekend pass, however; I was not one of them, due to the sergeant stopping when he got to me, and some others; we were unable to persuade him to do so. However, a week or so later, other soldiers and I were transferred to Fort Meade Army Base in Maryland. There I received a weekend pass. I left the camp and began hitchhiking to Washington, D.C., to see my mother, who lived on Hayes Street, North East. A cab driver picked me up, and upon giving him my destination, the driver stated that he could not take me there, but would drop me off at the

D.C. bus station. There I hailed a cab owned and driven by a black driver (Capital Cab), and was taken to my destination.

On Sunday evening, I got a ride to the bus terminal, and returned to Fort Meade. On Monday morning, I was ordered to appear before the company commander, who informed me that I was in the group to be discharged immediately.

I returned to my mother's house for a few days, and then went to South Carolina to visit my father and stepmother. I went around visiting as many relatives and friends as I wanted to, and finally my girlfriend and her family. After 3 weeks of being home, I returned to Washington, D.C. My family worked at the Statler Hotel in Washington, D.C. I applied for a job there and was hired as a "valet boy." That is what the personnel staff referred to me as in that job. I had been called "boy" many times in the past, and it did not bother me. The job was clean, not bad, and I made a lot of tips. I lived with my natural mother in Washington, D.C., and I was able to rent a room in her house for 20 dollars a month. I was able to accumulate enough money that made me comfortable in wanting to go to my home in Ellington, S.C, for a few days with my father. I had worked well over a year without a vacation. I asked the boss for time off, and he said no, so I hung up my little black velvet coat on the rack, collected my pay, and left. In a day or so, I left for home in South Carolina. After seeing my father, stepmother, and girlfriend, I returned to Washington, D.C., to my mother's house. A notice was sent to me at her home, directing me to the American Wholesale Mattress Factory, located on Fifth & V Street N.E., for an interview. I was hired on as a trainee making mattresses. The job did not require any technical skills; however, you had to pay strict attention as to how you were instructed to perform. I quickly became rather proficient in mastering the technique, so much so, I was elevated to a position of a tape edge machine operator. There were two machines of this type in the plant, and only one person was qualified to operate them. We had an hour for lunch, but in an effort to develop the necessary skill quickly to be an effective operator, I would eat my sandwich in half the time and immediately return to learning how to operate the machine, because it was rather tricky. On the head of it was something referred to as a shoe; you had to keep the tape and the cover in the shoe to close up mattress in making it a finished product. Once the tape slipped from under the shoe, you had to rip the tape off, and start all over again. An older man named Mr. Robinson told me to just keep trying because I could do it. He was efficient at operating the machine himself, because he was the only one who did it. One Friday evening about quitting time, the boss called me into his office, and said he saw me making progress at operating the tape edge machine. He then said he would come over that Monday morning to observe me operating the machine, and if he liked my proficiency, he would elevate me to the position immediately. I proved my proficiency, and he gave me the position. Six weeks later, I was considered as a master of the tape edge machine.

While working at the Statler Hotel in Washington, D.C., I had taken every examination given by the United States Civil Service Commission for which I believe I could possibly qualify. I received a notice that I qualified as a Mail Handler Distribution Clerk at the United States Naval Gun Factory, located in South West, Washington, D.C. I resigned from the mattress factory, and reported as scheduled to the U.S. Naval Gun Factory.

Theodore R. Coleman (Left) And
James Governor Coleman (Right)
Young Brothers In The South.

Brothers Forever

James G. Coleman and Theodore R. Coleman

Chapter 2

Employment at the U.S. Naval Gun Factory

I reported for duty at the U.S. Naval Gun Factory in October 1949. After orientation, I began my probationary period. After successfully completing my probation, I was assigned frequently as a fill in for a senior worker, who had an addictive health problem and would be absent for weeks at a time. I understood that the individual was nearing retirement and that I was most likely the person to succeed him. I had learned the job in great detail, and had not lost a single piece of mail while routing the office mail service. My immediate supervisor suggested that I apply for the job, and I did so. I served in the vacant position for 6 months while the personnel office was attempting to fill the position. They eventually filled the position with a white male from outside the confines of the Naval Factory. He was made the supervisor of the mail department, and I was identified as his assistant without any increase in pay. When I learned of this, I went to the personnel office to inquire why I was not selected for the job that I had been doing solely for nearly a year. I asked if something was wrong with my qualifications. I was told no and that the job had been filled by someone outside the agency.

I felt that race played a part in my not being selected for the position. I had no one to turn to, in order to address what I believed to be an action of race discrimination. After a few days of gaining my composure, I remembered the director of the Potomac River Naval Command. I handled his mail on a daily basis, especially the type that was identified with special binding and special address labels. I made certain that he got his mail every day, and on time. I was approached one day by two men who identified themselves by showing their official badges. They questioned me quite extensively about his mail. After they satisfied themselves of my answers, they told me that he was missing some of his mail that had been sent from various offices in the region. He told me one day that if he could be of any help to me in any way, just give him a call, or come to his office. I went to his office, and much to my surprise, he had just learned from his office staff that I had not gotten the job as lead mail handler in the mailroom. After a brief conversation in his office, he asked me if I would like to be transferred to the Supply Control Department as a supply clerk. My response was yes. He called the supervisor in the Industrial Relations Office and told him that I would be coming to the office to discuss the mailroom situation. He also told the person that he had a vacancy in his supply department.

I went to the Industrial Relations Office, and expressed my concerns. Afterwards, I was given a Form 57 to complete. Soon after, I was informed by the staff in the office of personnel to report to the Administrative Office of the Supply Distribution Center. My position was to receive requests for property and maintain a meticulous electronic filing system of everything within the command from nuts and bolts to battleships. The stock availability was immediately provided to his staff. Aside from the high priority type request, requests were filled throughout the command.

Although I was interested in the Fire Department, I really wanted to be a police officer. It appeared that police officers were in tune with the community. You would see them walking the beat and talking to people on the street, so I thought that would be a good thing to do in being a public servant. With that in mind, I started inquiring about it. Because racial discrimination was so much in evidence, I wanted to know exactly what I would encounter as a police officer. I was told that there were no black police officers riding anything in the District of Columbia, and that they all walked the beat. I found that to be true. What really got me was that for the first five years of employment, black officers had to walk the beat in order to be considered for any type of assignment.

My desire to become a public servant was enhanced. I looked at it from several perspectives. There is more than one reason for a person doing or not doing something. When I discovered the situation about walking the beat, the discrimination issue raised up in my mind. If we are talking about discrimination at a level of walking or riding, evidently there must be something else there that would prevent me from achieving in any other area.

In talking to members of my family, I asked what would be expected of me as a fireman, and I also asked about racial discrimination because coming from the south, we were always plagued with the element that was so profound in our way of life.

Since I worked at the Naval Gun Factory, an agency of the United States, government, and was subject to discrimination, I wondered how this would impact on the Fire Department. I did not know anybody else to ask in the fire department, I was trying to base my thinking on which department I would rather join. I selected the fire department and passed the written examination. I was disqualified upon taking the physical examination, because I did not meet the weight requirement of 145 pounds. I weighed 144 pounds. In the denial letter I received, I was informed that I could get a letter from my physician indicating that I had attained the required physical weight, and be restored to the roster for re-examination. I did that. Subsequently, I received a letter from the Civil Service Commission to retake my physical examination for the fire department. I passed the physical, and an appointment was open to me. I resigned from my position in the Naval Gun Factory. This process was required, because if I were not accepted for the fire department, I could return to my Naval Supply position. I accepted a position with the District of Columbia Fire Department, as a firefighter.

Chapter 3

Appointment as a Firefighter
District of Columbia Fire Department

When I was appointed to the Fire Department on January 5, 1953, I went to the Fire Alarm Headquarters located at 300 McMillan Drive and reported to the Lieutenant, who handled all new appointees coming in. His duties were to ensure that all recruits coming to the department were made aware of some of the conditions that he expected them to comply with. He talked with me, and said that he was happy to meet me, and said that he was known for having a photographic memory because he could remember everybody's name maybe even back 10 years or so. We had a casual talk and he gave me my probationary book and took me to meet the Fire Chief. At the time, I did not know the Fire Chief's name. However, he was sitting in a big chair, and was dressed in his official uniform with a white shirt and black tie. When I went in, I saluted him. I was used to saluting, having been in the military, and I felt meeting the highest-ranking officer in the fire department warranted a salute. I spoke to him, and he said, "Sit down boy." I sat down, and he asked, "Why do you think you could be a firefighter? Have you had any training for a firefighter?" I said, "No sir, I do not have any training." He asked, "What makes you think you could be a good firefighter?" I said, "Chief, I believe whatever training the fire department has to offer, I can learn it and be one of the best." He said, "You think so," I said, "Yes sir, that's what I believe." He said, "Do you know where you are going to be appointed?" I said, "I understand that I am going to be stationed at Engine 7, primarily because I live in that area." He said, "Did you select Engine 7?" I said, "No sir. I would like to be assigned to Engine 27." He said, "Do you know where Engine 27 is?" I said, "Yes sir, I know where it is." He said, "Do you know where Engine 7 is?"

27

I said, no sir, however, I understand it to be in the Southwest quadrant of the city. I don't know where Engine 7 is, I have never been in the Southwest area before." He said, "Well, why do you want to go to Engine 27?" I said, "Because I am familiar with that area and I lived close to Engine 27." He said, "All right, this is the kind of job we don't tolerate people being tardy. I'm warning you right now, if you are late for duty one time, you are terminated. It's just that simple. I can't tell you any other way. If you are late, you are terminated. We expect you to be at your duty station at the appropriate time every day. Most firefighters report to work as much as an hour before time to ensure time accountability. My only concern is that you be here on time. Six o'clock or 8 o'clock, do you understand that?" I said, "Yes sir." He then asked, "Do you have all your paperwork and your probationary book? I said, "Yes sir." He said, "Go over to the Uniform Board and get your uniform, and then report to Engine 7 where you will assume duty today. Did you come to assume duty today?" I said, "Yes sir, I'm ready to assume duty today."

I went to the Uniform Board on Maryland Avenue, N.E., and they gave me my uniform consisting of pants, shirts, boots, helmet, and running gear. In seeing other firefighters getting their uniforms issued, they were short and I wondered how they got in the department. I was "5' 8-1/2" and was right at the limit on the height requirement, and they were much shorter than the required height. I told a firefighter that I was denied entry into the fire department in 1950, and this is 1953. I told him that I could have been working here in 1950, if I weighed enough. He asked me what I meant, and he stated that a lot of things were happening in the fire department. They would turn down applicants for being pigeon toed, cross eyed, or anything in the world. He told me that there was a quota. They had so many blacks to be appointed, and they had to have somewhere to put them. I was interested in this conversation because it was all new to me.

There was a Captain in charge of the Uniform Board. Another firefighter was appointed at the same time and received his gear along with me. We conversed during that time, and departed by shaking hands. He went to Engine 26, an all white company, and I went to Engine 7, an all black company, one of the 4 all black stations.

Upon arriving at Engine 7 located at 347 K St., N.W., I walked in the door; some of the guys in the sitting room saw me scuffling with my things. A firefighter named Huscans jumped up and opened the door, and they all stopped playing cards to welcome me, the brand new firefighter. Someone said to me, "I guess you want to see the captain," and I told him yes. They all assisted in locating a place for me to put my gear until such time that I returned from seeing the captain in charge of the station. I went down front to see the captain. The firefighter named Huscans introduced me to the Captain, and the Captain said he wanted to give me an overview of what was to be expected of me as a firefighter. I gave him my Probationary Book, which was a requirement because the captain had specific information to put in the book such as short streets, court and alleys, 100 fire alarm boxes, 100 fireplugs. In my local alarm district, we had to memorize 100 fire alarm boxes and 400 hydrants. (There was no use in giving an individual all of that information because all of the probationary books were the same until you reached your assigned company.)

The Captain told me where we were going to begin and he showed me the watch desk and its functions. This is where all alarms are received from Communications, and all information requiring responses affecting this company's Communications Division. He indicated that as I learned my probation, I would know the difference between box alarm and local alarms. He referenced the watch desk again and said when an alarm was received at Engine 7, Truck 10 responded, you then hit the turn out bell. In summarizing a company response, he told me that I would be familiar with that because they line up every day by sounding the turn out bell. There were two strokes at 8 o'clock in the morning and 6 o'clock in the evening. That meant everybody lined up for roll call. He explained that the Captain would stand at the head of the lineup, and would check to see if everyone was there and his condition, and assign everyone their specific duties for the day. He said that I was to stand watch with someone during the day shift. After the cleaning of the firehouse and all apparatus had been completed for the day, he wanted me to stand watch with someone else. He stated that when he felt I was efficient enough to stand watch alone, he would let me do so because it was very important that we not miss any important runs. He said, in the mornings the first thing we do is clean up. After the 8 o'clock lineup, we cleaned up the firehouse and apparatus. The drivers were responsible for all rolling apparatus being cleaned and ready for immediate responses. All other firemen were required to assist in the remaining house cleanup, prior to the 10:00 a.m. drill time. Where we were assigned in the station determined where we were assigned to the small line on the fire truck with a little reel on it. Other day assignments were for brass and windows as well as changing of the hoses. If we were assigned upstairs, the three men upstairs would take care of the officer's quarters, bunkroom, bathroom, etc. He said other than that, my probation was my primary focus for one year. And, as I read and studied probation, I could have been dropped from the roll at any time during a period of unsatisfactory service, misconduct, or inefficiency.

The Captain explained that his duty was to examine me every month for 12 months. He indicated that he would take my book and ask me questions pertaining to its contents. At the end of a 3-month period, I would be directed to see the Battalion Commander, and after the next 3 months, directed to see him again. He stated that I could finish my probation in 9 months if I so desired. (The normal span for this is 12 months.)

The Captain pressed a button and asked a private to come down front, and escort me upstairs to show me the bunk room. It was all mapped out as to where everybody was to sit or sleep.

The officers' quarters were separated by a partition from the bunkroom where the firefighters slept. The bathroom was located at the other end of the bunkroom. A roster was hung by the watch desk indicating who slept in each bunk. The firefighter showed me my bunk, and said there were three areas that they worked on up here. He explained members of the truck and one member of the engine company cleaned upstairs. They cleaned, mopped, and made up the beds. All of the beds were supposed to be in line with blankets on them. Three men were responsible for this whole area. When the Captain came upstairs, everything was supposed to be in order. He also said that all of the shades were to be down at the same level. There would be no days that the house would not be washed out.

The firefighter and I were never off on the same days. There were different paddles, meaning days off. The day off paddle indicated when we were off. He told me he did not know what day I would be off, but whenever he was working, he would work with me to make sure I stayed on track. He said he would be asking me for my book from time-to-time to see how I was learning my probation, and I told him okay. After the tour of the house, we went back to the watch desk.

The Captain told me that there were many boxes to learn in the first month, 20 boxes and 10 hydrants. He asked the firefighter to show them to me. We had to see 7th Battalion Chief after 3 months for probation review. The first month we were required to know 20 boxes, so many fire hydrants, etc.

It was time to change the watchman at the watch desk, and the one that came down was firefighter Burton Johnson. He was the wagon driver of Engine 7. His job was to drive the wagon to fires everyday he worked. He had a 1937 Peter Parch Wagon, which was our fire apparatus at that time. It looked new because that was the way he kept it. I was so impressed with his ability to respond to fires that I wanted to emulate his ability to operate a wagon. Also, I admired one other driver named Jefferson Lewis, of Engine 7, Truck 10.

Firefighter Johnson sat down and asked me questions such as where was I from and where I lived, and made small talk about my children. I enjoyed the casual conversation and his interest. The Captain had told Huscans where to put my clothes on the fire truck. Later on, Huscans told me to leave my shoes in the house.

I did not forget what the other firefighter told me about leaving my shoes in the firehouse. We got a run one day, and I jumped out of my shoes. They would kid me about that, and they never forgot it. One day the bell hit box alarm Engine 7, Engine 13, Truck 10, and the guys started running. I ran around the corner and put my boots on. It is a funny thing to firefighters that, rookies look like rookies. It is a long time before they stop looking like rookies. When they get on the fireground, it just looks like they do not know what to do. We responded to the box alarm at 416, First Street, S.W. It was a false alarm, and we returned to the station. The Captain said to me, "You got your first run?" I said, "Yes sir." He asked, "What do you think?" I said, "I think I'm going to be alright Captain." He said, "I think you're going to be okay." That day, tour ended at 6 o'clock; about 4:30, my relief man came to relieve me from duty at the watch desk. The Captain told me that my relief man was there, and that I could leave.

Starting the next day, I had 4 days off—the first two of that current week and the first two of the next week. I did not know that until I came back to work, my tour would be 24 hours on a Saturday. I went home and my wife Uvaghn asked me how I liked the job. I told her that we had a little run. She asked me when I was supposed to go back to work, and I told her in 4 days. She was amazed. I told her that was the way it was.

When I went back to work, I worked the whole week including 24 hours on Saturday, and got off Sunday morning, returned Sunday a.m. for regular shift work. That worked out pretty good. The first month of my probation, we caught a few fires. The first fire I caught, there was a death. The first person that the Captain called was me. He said, "Coleman go out and get a blanket, we've got a dead body in here." Inwardly I thought I do not like being around dead people. I went out and got the blanket, and he asked, "Do you have your gloves on?" I said, "Yes sir." He said, "You and another person spread the blanket out, grab the body, place it on the blanket, and fold it up." We brought the body out and the next day my picture was in the newspaper.

For the first month, the Captain told me that he wanted to hear my probation. We sat down. He asked, "To be successful in the fire department, what must you do?" I answered that and all other questions down the line. He also asked about boxes, hydrants, short streets, and alleys for that month. I said to him, Captain, I don't want to appear to be out of focus but you can ask me the rest of it if you want to." He said, "What did you say?" I said, "If you're going to ask me for one month, you may as well ask me for two, because I know it." He said, "You learned two months in one month? I guess you are going to try finishing in nine months." I said, "Yes sir, I will finish in nine months." He gave me a form called Form 121 to fill out and put in the basket for the Chief. The guys asked me what had I done to the Captain. I asked them what they meant and they said they never heard of anyone getting an "excellent" during their probation. They said that the Captain and I must have had something going. I told them they should be ashamed, and that I was only with the Captain for an hour. They still said they never saw anyone get an "excellent."

For the next two months, I got an "excellent." That is when I became a very knowledgeable firefighter. I then went to see the 7th Battalion Chief. The Battalion Chief wanted me to ask myself questions, and give the answers. I knew some of the questions, those in the order book, the manual, and the rules and regulations. It was not a normal procedure to have you ask yourself the questions and answer them.

I went back to quarters with a "very good" and I told the guys what happened. They could not believe that the Chief instructed me to ask myself the questions. They talked about that for a while that day. The Chief said that he knew from looking at the Captain's report that I knew my probation. He told me that I did not know the questions. And, in order for me to get "excellent," I had to know the questions and answers. I thought to myself they never told me of anything like that. The Captain said to me, "The Chief couldn't give you an "excellent." I said, "No, he couldn't give me excellent because I didn't know the questions." He said, "I have never heard of that before, but if that's what the Chief wants, he's got to get it."

The third month when I went to see the 7th Battalion Chief, I knew all the questions and answers for 3 months. The Captain told me in the beginning I was not to play any ping-pong, do any car washing, or anything else until such time that he felt that I knew my probation. After the first month, the Captain told me I could play ping-pong, wash cars, or anything else. He just told me not to try and do too much on the fireground because he noticed at the fire we had that I was all up in there. He wanted me to stay close to the other men. He said that fire was only on one floor, but if we had to go into a place at night, they may not be able to locate me. So, he reiterated that I was to stay close to the other men. He mentioned that he knew that I wanted to get in the fire. The old firefighters coming from Engine 27 and some other places believed in getting in there and knocking the fire down. When they started putting you on the line and leaning on your shoulder, they would say, "Get in there, get in there." There was nowhere to go with a 200-pound man pushing on your shoulder.

After 9 months, the Chief was going on leave and the Captain asked me if I wanted to go and see the 7th Battalion Chief, and I told him that I was ready. I went to see the Chief, and he asked me if I wanted to finish the 9-month probation period, and I gave him my entire book of questions and answers. The Chief called the Deputy Chief of the Fire Suppression Division, and told him that he had me with him, and wanted to know if his schedule would permit him to see me in order to have me complete my probation in 9 months, and he told him ok. When I got there, the Chief looked at my probation book. It was a new book that I gave him, and one I kept in my pocket. But in the process of putting it together, I was about 20 boxes short which I left out. He never knew it, because he did not ask any probation questions. He just asked me about how I liked the Fire Department, and said he noticed that the Captain gave me excellent on everything in quarters, in the house, and on the fireground. He said that he could not expect anything more, and there was no use in him asking me anything else. He did ask me, if I had any idea as to what the line pressure was on the gas stove, I told him I did not know. He said 3 pounds.

I had one more river to cross. Having passed my probational requirements, I was ordered to attend the Training Academy. The Probationary Book made it clear that you had to pass at the Training Academy, because if you did not, you could be dismissed.

Chapter 4

Mandatory Training at the Fire Academy

This was a requirement for new recruits in the Fire Department at a point and time in their progression, as a well-informed firefighter. On my first day, I met the assigned staff, which consisted of a Deputy Fire Chief, two Captains, and an aide to the Deputy Fire Chief. We were made aware of the conditions associated with our 45-day tenure at the academy exclusive of Saturdays and Sundays. We were told that we were expected to be at the Training Academy at 8 o'clock in the morning until 4:30 in the afternoon. Also, we were told that we had a job to do, and the officers at the academy were going to do everything in their power to ensure that when we left the Training Academy, we would know what it was like to be a firefighter in the District of Columbia Fire Department.

It was made unequivocally clear that playing in the training yard would not be tolerated. The training yard was designed primarily for training, and the classrooms were identified for study. We were told in the morning that we would line up, and we hoist the American Flag on the flagpole at the Training Academy. I was designated for the job, because I was next to the shortest man in the class. We were told that anytime there was a third alarm sounded, we were to report to the Training Academy.

The training got underway, and we received our training manuals, our studies, homework, and things that were pertaining to the training curriculum. The training began the first thing in the morning. Safety nets were put in place should there be an accident while climbing the pompier ladder. When our training began, the officers would go through every chapter of the training manual, such as:

- Line up for receiving official visitors
- Tools and appliances
- Special equipment and personal equipment
- Rope and ladder
- Fire hydrants and hose
- Flammable liquid extinguishing agents and equipment
- Heavy-duty equipment
- Rescue work and forcible entry
- Ventilation
- Overhaul and salvage
- Automatic sprinklers and standpipes
- Refrigeration
- Fireboat and lifting devices
- Minimum equipment standards
- Schedule of drill evolutions

During that period, hurricane Hazel struck the area. The hurricane was identified as one of the most destructive that we ever had in the District of Columbia. With the devastating effect of "Hazel," a third alarm was sounded, and we reported to the training academy as directed. The Captain at the training academy was in charge of the made up units that responded. A piece of apparatus was made available so that we were ready for any emergencies in that area.

We received information from the Communications Division to respond to the 3500 block of Stanton Road in Southeast, D.C., for a house that the wind had blown the top off. We responded to the alarm, but it was nothing we could do.

We reported in service. On our way back to the Training Academy, we were dispatched to respond to the 400 block of M St. S.W., D.C., which was in my local alarm district. I was also selected to perform as the wagon driver during that period. When we reached the 400 block of M St., it appeared that all hell had broken loose in that area. The bricks from the chimneys were flying around like hail. It was amazing that some of us were not injured during that time. There were no fires to respond to. However, at another time there was a third alarm sounded at the Peoples Drug Store at Logan Circle, N.W., and members of the Training Academy responded to the drug store fire. Upon arriving, there was nothing for us to do but assist in getting some of the merchandise out of the basement of the store.

The members of the Training Academy were very competitive. Every Friday when we had a test, everyone would want to know who made the highest test score. Our class consisted of 12 members, which was comprised of 10 white firefighters and 2 black firefighters. Calvin Watson was black, and always on top, and I would be somewhere in between, two or three slots from the top. We always had the competitive spirit. As firefighters, we were always eager to learn how to be the best firefighters by paying attention to all the training that was made available to us at the time. We had very little time for playing. Everything was scheduled in such order, as to start at 8 o'clock, 12 o'clock to lunch, and at 1 o'clock, we were back on the job. This was the order of the day for 45 days. At the end of the 45 days, we had what was called a qualifying day, in reference to climbing the pompier ladder. There was one individual who beat me out on the pompier ladder, because he did not follow the instructions that were given by the training instructor. His name was "Horn" from Engine 15. It was a ladder made like a stick. It is a long stick with a hook on the end with rungs going down the middle of it. We would hang it in the window above us, and then climb up. When we reached the windowsill, we would reach down with our left hand, pull it up, hang it to the next windowsill, go up, and continue until we reached the top floor. The Training Academy tower had six floors. I did it in 58 seconds. I believe "Horn" did it in 52 seconds. I had no problem at all with the aerial ladder, because we had a truck company in the house where I was assigned. Truck 10 was in our quarters and I had been doing that every since I was appointed. Every week when we had a drill, I had to climb the ladder, hook in, and lay back 100 feet in the air. There was a pompier belt that only clicked one way that locked so we could work on the ladder with both hands. If we were on a regular extension ladder, we used a leg lock so that our hands would be free.

One or two guys dropped out of the training class because they could not pass the test. In addition to the written test, we had to identify equipment by putting our hand on each piece of equipment, carry it into the rating room, explain to the Captain how it was used, and say exactly what it was for. This was the verbal part of our training. There was not much excitement at the Training Academy, because it was programmed in such a way that everyone knew what was expected of everybody, and what was expected from the instructors. After our training curriculum was over, we received our certificates and returned to our respective units. My class number was 338.

At this time you could see the divisiveness in the way we were treated. A black firefighter inspired me to some degree because he withstood a lot of trauma in being a member of one of the first group of blacks to be assigned to a white company. They are the ones who I considered as pioneers, because every house I went to, there had already been black firefighters. So, I was not the first one to be assigned. The black firemen would come to work, sit down and read the newspaper, and not have much to say. They were the ones that were sent to these companies. I remained at Engine 7, and I was detailed quite often to Engine 8, an all white company, to fill in for a firefighter who had gone hunting while on annual leave and was killed. The tour that I assumed was a 24-hour shift. Since they did not have a bed for black firefighters, I was returned to Engine 7. My departure left them with a four-man shift. Engine 7 had three extra staff on duty. What was happening at that time, especially at night, was white firefighters would be denied annual leave because they had to have five men on the firetruck. Engine 7 could have three or four men too many, but they could not send one of those men to Engine 28, to let a white firefighter have leave. So, it was a problem for both white and black firefighters. During the day when they had too many, they would put us out to inspect buildings, apartment houses, boiler rooms, and fire extinguishers in schools. We would go to the Battalion Chief's office, and he would give us a route book that had inspection routes, apartment houses, etc. There was to be an extinguisher on every floor, and then on some floors there was to be a CO_2 fire extinguisher, all having tags on them. An inspector would write all these things down.

(See Appendix A for concise history of the District of Columbia Fire Department.)

Chapter 5

Incidents of Discrimination and Racism in the D.C. Fire Department

Racism is defined as a program or practice of racial discrimination, which is a side product called segregation. Out of this comes persecution and domination based on racial bias. On the other hand, prejudice is defined as a judgment or opinion formed before the facts are known, a preconceived idea, favorable or unfavorable. Or you could say, it is a judgment or opinion held in disregard of facts that contradict it; in other words, unreasonable bias in both cases.

It was only a misnomer that black firefighters were fearful of fighting fires. Perhaps this was the real reason that integration was opposed. It is my intent to establish the veracity of truth concerning this misnomer, and to share my experience relative to factual incidents concerning the mentality of white members. The mentality of discrimination caused much disruption and grief for all black members.

My very first assignment in the D.C. Fire Department was at Engine 7, an all black company, and one of the busiest companies in the department. I readily discovered that white rank and file members distinguished themselves as superior to black members. They performed the regular run-of-the mill type fireground, duties and were frequently honored by top management within the department, as well as local government and other private sector groups, for heroic acts of bravery.

More was always expected out of the black members. Older members of this ethnic group took the liberty of ensuring that all new black members were fully aware of firefighting protocols and personal safety measures, since we did not have the luxury of to-day's modern equipment. Our job on the fireground was much more strenuous. We climbed ladders, fought heavy fires under extreme adverse conditions, both from within and without all type of dwell-ings and structures. Also, we saved many lives. Did we get any rec-ognition? No, only that "it was part of our job." If a report was writ-ten commending our efforts, it was mysteriously destroyed before reaching top management.

Although I was subjected to much racism and discrimination, I managed to advance to the rank of Deputy Fire Chief and was placed in charge of the Training Academy. I also served in the capacity of the Chairman of the Advisory Board of Awards. In this capacity, I implemented guidelines to recognize all firefighters who performed their duties on the fireground above and beyond the call of duty. This recognition entailed the receipt of the gold medal award or other similar like awards.

During this era it was either an unwritten policy or just plain perception that all white officers be temporarily assigned at Engine 7, Truck 10. Thus, two white officers, Captain and Lieutenant, were subsequently assigned. They made some effort to improve cama-raderie between whites and blacks. The Lieutenant openly stated that he felt the members of this unit far exceeded members of other companies he had been assigned to, because of their dedication and commitment. He also stated that he was proud to be affiliated with this unit.

While working as part of a team at a fireground incident, a black firefighter of this unit performed his duties in an exemplary manner and was subsequently written up by the Captain for proper recog-nition. The Advisory Board of Awards agreed that the actions of this firefighter warranted the "Gold Medal Award." White firefight-ers collaborated among themselves stating that the report contained erroneous information. From this, white officers became involved. The irony here is that they did not want a black to receive the "Gold Medal Award," which was the highest recognition one could re-ceive. Nonetheless, when I became Chairman of the Board, I took a firm stand in upholding the decision. It was a tedious task, but we were able to sustain the initial recommendation of his Commanding Officer. Had we failed, the award recipient would have been white.

In early October 1956, the climate was very, very dry. No rain fell for several weeks, and the soil was extremely hard. Georgetown University planned to have a football game on their field. Because of the condition of the soil, the University contacted the Fire Department to have a company water it down. This request was granted by the Fire Chief. Now, in the general vicinity of the University, the first due companies to that area were Engines 23, 29, 16, 1, 5, 20, and Truck 5. Instead of sending one of these units, the Fire Chief opted to draft members from Engine 7 to handle this detail. The units previously mentioned did not fight nearly the number of fires that Engine 7 did, nor was there a comparison in performance. Engine 7's firefighting expertise took no precedence over material details when it came to alleviating white members of undesirable assignments. Nevertheless, myself and other members of Engine 7 proceeded to the detail. It began raining like you would not believe, if you were not there. This detail was subsequently cancelled by the Fire Chief.

In another incident, there must have been at least 5 feet of snow on the ground and the weather was extremely cold. Many low-income people lived in wood frame houses and used pot bellied stoves and kerosene heaters to heat their homes. These methods of heating were most dangerous. Engine 7 responded to numerous fires in poor neighborhoods during the cold season. The company ran practically all day and all night to fight heavy fire and intense smoke. Many lives were lost due to the heating mechanisms used. It seemed each response was worse than the first, and, whenever we returned to quarters we were totally exhausted. Notwithstanding fighting fire, each company was responsible for the replacement of its own hose and equipment, which meant additional work.

Later, on this snowy cold night, Engine 7 had many responses. Engine 25, an all white company, had a relatively small fire at Bolling Air Force Base. When they finished, they left the scene leaving behind much of their hose. Engine 7, the all black company, was placed out of service, to be transferred to Bolling Air Force Base to retrieve Engine 25's hose and deliver it to them. This was a clear act of racial prejudice, which kept hostility growing.

During the era of integration, blacks were periodically detailed to white firehouses. The tour of duty referred to was a 24-hour shift; details ended either at 8:00 p.m. to 8:00 a.m. Whether you stayed for the next tour of duty depended on whether there was a bed marked "black." If there was no such bed identified, detailed black firefighters were returned to their respective all black company. A vacancy was created at Engine 8, an all white company, by the death of a fellow firefighter. I was then detailed there to fill the void. Since this company did not have a bed marked "black," which would have been assigned to me at 6:00 pm, the company was reduced from five to four men to eliminate my being there 24 hours. Thus, I was ordered to return to Engine 7.

Requesting annual leave became an enormous problem for white members. Requests for leave were denied when there were extra men in black companies and there was a shortage of men in white companies. Leave requests were denied simply because they did not want to use a black firefighter as a replacement in an all white company. This also placed a heavy burden on black firefighters not knowing what would be required of them next. They wanted very much to be efficient and effective, to say the least. Blacks wanted everything to be perfect and intact when they were on duty, because the system dictated it to be that way. Whites showed little concern.

During the renovation of the South West section of the city, near the waterfront, many houses were torn down by a construction company. One day the weather was bitter cold, so the workers built small fires to keep warm. A fire got started on some debris, resulting from live coals. We responded to the fire location. We had a black sergeant that night. For all intent and purposes, we will call him "Little Willie." Sgt. Little Willie and the firefighters stayed out there in sub zero weather from 12:30 a.m. until 4:45 a.m., throwing water in what was left of this vacant house. Anybody with good common sense would have tried to protect his men from that extreme cold weather, when there was no fire, and nothing that could catch on fire. There was nothing in the area; the house was gone. The construction workers were there getting the mortar off the brick, so that the bricks could be used in other construction. We stayed there for 5 hours, because this sergeant wanted to be so efficient.

The Battalion Fire Chief learned of this behavior, and ordered the company to return to the quarters. It was ridiculous, but this is the kind of Fire Department where there was so much pressure put on the black officers that we did not stand a chance. They would work us nearly to death. We would begin working as soon as we came through the door, especially during the warm weather. We worked on apparatus every day, all day; nowhere in this agency's history can you find anybody taking 10 to 15 days to clean apparatus all day for inspection. We got Lysol and brushed down all grease and oil. If we got runs that night, or if we did not get any runs at night, the next day all of the hose would have to be taken off the apparatus, and we would start cleaning again. White firefighters were given advance notice of the inspection date, and they cleaned the apparatus at their leisure, and they were prepared for inspection.

I blame the system; some of the black officers were good people, but the system dictated their behavior when they were put in charge of a black unit. It had to be 100 percent pure, and that created an enormous problem in the black firehouses. You could easily see the contrast when we had white officers. The white officers did not seem to be concerned with little details as the black officers. The thing that was appalling to me was when we had total integration, the concept was to do things right. When black officers were in charge of white firefighters, they did a complete 180-degree turn around. They were not as demanding in making sure that everything was done right. It was so pathetic how those black officers would run rough shod on the black firefighters. I believe that fear of being ridiculed was the dominant factor as to the reason for their behavior. I hasten to say that all black officers were not that way. Some had their own way of operating.

We speak of things being fair. If you were to ask yourself the question, who determines what is fair, I believe you would answer that it is the people in charge that determines what is fair. There were enough black firefighters and officers in the D.C. Fire Department to diminish racial disparities in the working category of the department.

When blacks were privy to discriminatory actions involving a fellow black member, nothing was ever said for fear of being ridiculed. And yet, away from the work place, they would discuss these actions among themselves and with others. This provided no rectification. They needed to speak out, and be heard. In a point of contrast, white firefighters and officers who were up for promotion were given tips to stay in line. I remember an incident involving a white Captain, who was a victim of alcohol abuse. Black fire personnel assigned to his company witnessed him being inebriated to the point that he could not respond to scenes of emergencies. On one occasion, this officer came to the firehouse while he was off duty and in civilian clothes. He was so obnoxious, I directed a Lieutenant to put him out of the firehouse. I was a Battalion Chief at that time, and I did not want to go out there, to do a Captain or Lieutenant's work. The Lieutenant was the ranking officer, so he put him out.

This Captain was at the top of the list for promotion. He was tipped off by a high-ranking white officer to improve his demeanor, or he would otherwise not be promoted. He stayed free of alcohol long enough to be promoted to the rank of Battalion Fire Chief, then it was right back to the bottle. Although blacks were cognizant of his addiction, still nothing was ever mentioned.

(For a historical example of organizational discrimination and racism, see Appendix B: Seminole Indian Scouts.)

Chapter 6

Black Firefighters Detail Assignment

In 1954, black firefighters were being detailed to all white companies. They were being transferred as well. I was detailed as well. The sleeping quarters of firefighters were separate from the officer's quarters. In order to maintain segregation of the races, the white officers gave up their sleeping quarters to the black firefighters, who slept in the all white bunkroom at Engine 5.

I was detailed to Engine 5. There was a black firefighter, who had been there for 6 or 7 months. He was restricted from using the kitchen, because he was told he did not own anything in the kitchen. He had to go out for his meals or bring them from home. He could not as much as heat up a can of soup for a snack. He was there by himself and had no one to talk to. This is what is known as the silent treatment. The one that received the worst silent treatment later became an Assistant Fire Chief (now retired). He was assigned to an all white unit on Wisconsin Avenue, Engine 28. He took the examination for sergeant and ranked number one. I would kid him about the silent treatment. He told me one time that he did not have to worry about silent treatment anymore. The Captain gave two strokes on the house bell, and I received my assignment on this particular morning, where I was supposed to be riding on the engine. I returned to the lounge area and joined firefighter Nelson. We began watching the "Joe Louis Story." He mentioned to me that we were having a good day, and I asked him what he meant. He stated that we were watching television. I told him that we should be able to watch television anywhere, and he told me, "Not here." He was a rookie in the department. I told him no one was going to mess with the television that day. He said, "Wait and see."

The white firefighters would turn the television desk around to favor their ability to see the set clearly, and away from the black firefighters. It wasn't long before a white firefighter came up and started to cut the television off. I responded to him rather aggressively, and said a couple of foul words. The captain came in and I started to tell him what was going on. The captain told me not to curse. I told him that the guy wanted to turn the television off and that I understood this had been happening here, and nobody was doing anything about it, and that was ridiculous. I told the Captain that he had just come from Engine 7, and he knew we had a good relationship there. Nobody turned the television off. If they didn't want to watch television with anyone else, they would go outside or do something else, go anywhere, but not turn it off. I told the Captain that Article VI, Section 5, of the Rules and Regulations clearly stated that whatever property was in the firehouse belonged to me as well as to him, whether it's on the property card or not. That guy knew absolutely nothing about the property card. Everything in the firehouse was recorded on the property card, or it's supposed to be. That is how they would lose things all over the place and could not find it, because the property was not recorded. When we came back into the firehouse from a fire run, we would check our tools and appliances to ensure that we had all of our equipment. The Captain told him to leave the television alone, because I was right about Article VI, Section 5, of the Rules and Regulation. The guy said, "He doesn't have anything in it Captain; we bought it." The Captain said, "It doesn't matter; it's in the firehouse."

That is the only problem I had, other than at Engine 10. I had a slight problem with the wagon driver. He was supposed to clean the wagon; that was his job. Take it out front or either in the firehouse, get a pan or bucket, and clean under the fenders, to get all the dirt and grit off. The pumper man would take care of his, and the truck drivers would take care of theirs, but this particular wagon driver failed to prepare his wagon for immediate use. His actions were completely ignored by the ranking officer. For example, he would let him get away with all kinds of things. As a result, other firefighters became inefficient. One day we received a run, prior to the apparatus being checked by the wagon driver. Sometimes the centrifugal pumps would get air locks in them, and if the air was not relieved, it wouldn't pump water. That's why we had primers on the pumps to get all of the air out of them, to ensure that it was ready for service. When they came back to the firestation from the Training Academy, the driver failed to prepare his wagon for emergency use by not getting the air out of the pump. They couldn't get water until the pumper let the water into the intake side of the wagon, causing the air to be expelled from the pump before we could get water to knock the fire down. Before then, we had to get water from the tank, carried on the apparatus. It would last for only a few minutes using a one and one half inch line. If there was any air in the pump, we couldn't get the water out of the tank. We called it "air lock." He said to me, "Why didn't you check the wagon?" I asked him what he meant and why I should have checked the wagon, since I was not the driver. He was the driver and that was his job. He said, "I'll tell you what, if we get another run, don't get into the wagon to drive back." This is driver training for firefighters. I asked him what he meant, and he just repeated that if we had another run I was not to get in the wagon to drive back. I asked him if he was an officer. He said no, but that he was the wagon driver, and "if you can't check the pump to see if there is air in it ...." I stopped him at this point, and told him that I was not going to discuss it with him. I told him that if we did get a run he could bet his bottom dollar that I would drive the wagon back to the firehouse. We did get a run, and I got in the wagon. The driver came and opened the door, and I asked him what he was looking for. He said, "I thought I told you...." I told him, "You better get away from around me before something get started." The ranking officer asked what was happening, and I told him nothing was happening with me, but he should take care of the wagon driver, because he was getting ready to create a problem for you, himself, and me. The driver got on the back step. When we arrived back to quarters, he said, "I didn't like what you said a while ago, Coleman." I said, "You didn't?" He said, "No." I said, "Okay, you didn't like it, right?" He said, "Right." I asked him, "What are you going to do about it?" Everyone saw a little feud coming up. The ranking officer was an ineffective officer in his leadership. He asked me what was wrong with his wagon driver. I told him that he should instruct the driver on what his responsibilities were as a wagon driver and what was expected of him. I told him that I had noticed the other day when you told him to watch out for Engine 8, whose running pattern was the same as Engine 10, in the same location to prevent a collision, the driver stopped the truck and asked if you wanted to drive. He said he would never drive nothing again. I heard him. Then I told him that he took the driver upstairs for counseling, but did nothing to him. I told him that he was supposed to put him on the back step, and that he was supposed to be in charge of the company. The driver was not in charge of anything. I spoke off the record to the ranking officer about the incident involving the driver not cleaning the wagon, nor his workstation, and the whole side of the apparatus engine house. Other personnel refused to do their assignments.

I had a talk with the Chief about not letting me act as Sergeant. It was my time to act. After successfully passing an examination, we were placed on a list to act in a position of authority when regular officers were absent. It was at least 6 months before I was first assigned to Engine 10, wherein the captain asked me if I was an Acting Sergeant, and I told him, "Yes." This wagon driver wasn't even on the list, but they were allowing him to act until I called the Chiefs and asked them to straighten it out. They did. The next evening, I was in charge of the company. While going across Benning Road, I told the wagon driver to back it down. He asked me what I meant, and I told him that he was driving too fast and to back it down. That is exactly what I meant. His response was of a very physical nature in facial expression.

We responded to an accident and I told the driver that we needed to get on the other side of the avenue, from east side to west side of Kenilworth Avenue, N.E. I told him to take the wooden ladder off the pumper, and put the hose up on it. He told someone, "He makes me do everything when he's in charge." As time passed, he began talking to me quite frequently, and he turned out to be an efficient firefighter. As an Acting Sergeant, we were qualified to be alternates in the Ambulance Service.

I was detailed frequently to units 5 and 7 in the first battalion. Being an Acting Sergeant meant that I had passed the written competitive examination, and was qualified to perform in the Ambulance Service. There were 75 firefighters taking the examination. I was not in the top 10; I was about 23 on the list. I made sergeant from being there at Engine 13, Truck 10. Truck 10 was the most active house in the fire department. We ran with everybody, for example, engines 1, 3, 6, 8, 18, 26, 27, and 30. Anytime one of these numbers was on initial response, you could be sure Truck 10 would be there. They would call out: box alarm Engines 3, 12, and Truck

10. Our Communications Division would respond by saying Engines 19, 27, Truck 10, Engine 30, 19, and Truck 10, Engine 30, 27, and Truck 10. Engine 30 and Truck 10 seemed to run the most frequent. From Engine 13, Truck 10, I went to Engine 11, Truck 6, as a newly promoted Sergeant.

Chapter 7

Promotion to the Rank of Sergeant

I was assigned to Engine 11, Truck 6, as a newly promoted Sergeant. One factor was generally understood as a new sergeant in a different firehouse: You could expect to be challenged in all areas of your responsibility.

I will never forget my first day at Engine 11, Truck 6. The guys wanted to test me. They were in the same house where we had a run. The way it was set up, if you received a run, say 14th Street was on the west side and 13th Street was on the east side of Park Road. If Truck 6 were due second, they would respond to the rear of the building. If Engine 11 were due first going out of 14th Street, then Truck 6 would follow in the rear of Engine 11.

The wagon driver was getting too close to the truck. The truck had ladders hanging over the rear. I instructed the driver to back it down, and not get too close to the truck. He backed it down. We were going to Flower Street, N.W. Flower Street was way out there and we had a good ways to go out New Hampshire Avenue. He wanted to put me to a little test, I believe, and followed too close behind the truck again. I told him to pull it over to the side, and get on the back step. I directed another driver to continue on. When we reached the location, the Chief noticed that the wagon driver was on the back step. The Battalion Chief knew all of the drivers, especially the first driver. They knew who should be driving the wagon whenever they were working. The Chief wanted to know why. He didn't ask me, but he asked the driver what was happening. No one knew anything yet because we had not returned to quarters. When we returned to quarters, the Chief called to talk to the Lieutenant. The Captain was off, and I was taking his place that day. Nobody knew about a driver replacement. I just took him out. It wasn't a big thing with me. I told him not to pull up too close, but he continued; so he was out, and I replaced him with another driver.

The Lieutenant asked to speak with me upstairs. We went upstairs, and I was asked why I removed the driver for another driver. I told him that the driver was following the truck too close going out New Hampshire Avenue, and I was always concerned about safety. And I told him not to do that. He did it again, and I took him off the driver position, and assigned him to the back of the wagon. The Lieutenant said it would be handled. He went downstairs and called a lineup. He told the men that there was a situation that needed taking care of. He told the driver of that wagon to step out, and told him that he understood that the sergeant told him to back it down, that he was driving too close to the truck in front. The driver told him that was right. The Lieutenant told him that he understood that he did it again. The driver told the Lieutenant that he wasn't paying a lot of attention. The Lieutenant said to him he didn't want to know about paying attention, he just wanted to know who did it. The driver told him that he did. The Lieutenant told him to return to the lineup.

The Lieutenant said, "Listen up everybody. I see there is a problem as related to Sergeant Coleman's authority. There is no mystery about who is in charge of Engine 11, Truck 6, when the Captain or myself is not here. When the Captain or myself is not here, Sergeant Coleman is in charge of the house or whichever unit he is responsible for." He said that he wanted to make it clear that he wasn't having any problems at Engine 11, Truck 6. "When the Sergeant gives an order to do something, and you fail to comply, he will relieve you from duty, and I will concur." He said that he had been a Sergeant, and he knew how it was when they put a new Sergeant through a test. Also, he told them that he had heard about me as an Acting Sergeant, and knew how serious I was about responsibilities. If they wanted to test me, he told them to go ahead, but they could rest assured that I would take appropriate action against them. When they were relieved from duty, the Fire Chief would determine the penalty based on the severity of the infraction. If they understood that, there wouldn't be a problem. If there was someone in the lineup who did not understand, the time for them to say something was right now. That was it, and that's all it takes to manage a company effectively in any fire station. Any fire station that blacks went to, when the captain fails to act responsibly, it caused problems to occur between the races.

The officers and men assigned to Engine 11, Truck 6, was very enthusiastic. They liked being assigned to the line, because they wanted to get the fire position along with the officer that was getting in there. The men were so proud of Truck 6, they named it the "Big Stick." I instructed the driver to display the "Big Stick" when responding to a fire position. I wanted to see the "Big Stick" in the air. A Lieutenant was assigned to Truck 6, and a Captain was assigned to Engine 11. We had a tremendous fire suppression force.

As a Sergeant, I was periodically detailed to other places. The Captain, or the Acting Battalion Chief, was close to getting promoted, and he stayed on the road quite often. This meant that I had to fill in his slot while he was away. One day, I was detailed to Engine 22, and we were all standing out front in the cool of the evening watching the people pass by. Salesmen would always approach us and try to sell us something, because they knew firefighters would buy. A salesman stopped and asked me if I wanted to buy a watch. I was always interested in jewelry. I took a course in watch making, and I was known for having a jeweler's eye. I could look at a piece of jewelry and determine what it was and the quality. The guy supposedly showed me a Longine watch, worth $300 dollars that he would sell me for much less. He opened the box and the watch was beautiful. The price tag had $375 dollars on it. I told the guy that I only had about $10 dollars and he said he would let me have it for $15 dollars. The firefighters watched as I gave the guy $15 dollars. He showed me a business card that was attached to his coat and left. Everybody wanted to see the watch. As I took it out of the case, I noticed it was extremely light. At that point, I wondered exactly what had I bought. We examined the watch with a magnifying glass, all to find out that it actually was not a Longine, but a "longune." Everybody knew that I had been ripped off with a hustler's watch. This transaction spread like wild fire. The Sergeant bought a hustler's watch. That is all you heard. For the next 2 years, I was reminded about the jeweler's eye.

Chapter 8

Promotion to the Rank of Lieutenant

I was promoted to Lieutenant. My new assignment was to Engine 30, Truck 17. When the order came down, it was customary for the promoted to give a promotional dinner for the company. Usually, dinners were similar to an everyday meal, which consisted of ham, boiled potatoes, and carrots that were all thrown in a pot. They would cut a head of cabbage in quarters, a piece for each firefighter. The meal for my dinner was different. We had roast steak, mashed potatoes, gravy, string beans, asparagus, other side dishes, garden salad, ice cream, and cake. The food was procured from a local market at my expense and cooked by Firefighters. A Captain was acting as the 4th Battalion Chief that night. He attended the festive celebration. We were lucky that night, having only one run that turned out to be an incident of more smoke than fire.

I was roasted at the dinner. They talked about how they enjoyed working with me as a sergeant, and they mentioned a few things that had occurred, especially responding to fires on Park Road. They also talked about my leadership and interpersonal skills in dealing with the men while I was in charge of the company.

The night of my promotion dinner, the guys had a box wrapped with the most beautiful ribbon you ever did see. The truck driver said the members of Engine 11 really appreciated my leadership while I was assigned there, and they wanted to give me a little gift to remember them by, once I was transferred to my new position. The truck driver was one who always pulled stunts and they all knew it. Sometimes his stunts made them angry. I figured he was going to pull something on me, but he acted so sincere that I started having mixed emotions. He continued my roast by making mention of a few incidents that had occurred in the company. We all laughed, and had a great time. When he gave me the box, I looked at it for a while, then I made my speech. The atmosphere had changed. It lacked the sincerity that had been there. The box contained a picture that was drawn by a cartoonist firefighter of me and the jewelry hustler standing in front of the firestation conducting business. The picture was signed by everyone. My promotional dinner was a great event.

Subsequently, I was assigned to Truck 16 from Truck 6, because there were more serious runs, but the Battalion Chief at Truck 17 had requested that I be assigned there. I submitted a memorandum to the Fire Chief requesting to go back to Truck 16, when a vacancy occurred. So, my tenure as a Lieutenant at Truck 16 was not long, 6 to 8 months at the most. Truck 17 was different; the men there were less aggressive than those of Truck 6, mainly because of the location of the companies. The district of Truck 6 had apartment buildings, high rises, and one or two single-family dwellings. There was a variety of dwellings in this district. On the other hand, Truck 17's district had a lot of one-story family dwellings, and only one or two six or seven-story buildings. We did not use the aerial ladder quite as much as we did at Truck 6, but we used the ground ladders a lot. It was also different because Engine 30, Truck 17, was the busiest company in the District of Columbia.

At Engine 16, the Captain and I had worked together on previous occasions, but not with my being in command of the company. The first thing I did when I arrived there was to ask the Captain not to allow the men to put a wedge between us. It was very important that we establish that kind of camaraderie to manage the company effectively. Once firefighters felt the company officers were divided, there were immediate problems and the Captain agreed. He mentioned that when he was a Lieutenant, the Captain of the station allowed him to study while he did all the administrative paperwork of the company, and that he would do the same thing for me. He did just that. He handled typing, reports, and time and attendance sheets for both units. He brought these to me for review. I would be upstairs studying and preparing for the examination and would read his reports for concurrence prior to submission.

I did not think about the racial makeup of the company, there were, however, three or four blacks on each shift, and no one complained. I assumed that things were running smoothly. Then, one day we were talking about giving a Lieutenant a retirement party, and I noticed that they could not get it together. Another firefighter and myself, decided to handle it. I decided to take care of the turkey, dressing, gravy, and the potatoes, and he would take care of the rolls, string beans, greens, etc. I actually prepared the dinner. In doing this, I wanted to set an example in the firestation for being able to work with all assigned members, and doing things as a group. Each company is different, depending on the officers; I wanted the same kind of harmony that existed at Engine 11. Sometimes new officers in a company, especially black members, experienced a period of rejection by the white members. That was the reason I wanted to ensure that we had harmony, notwithstanding the fact that I also liked to cook. Since I was partial to my own cooking, I did a lot of it. Everybody participated, and it was just that simple. We seemed to be getting along very well.

One morning a truck driver asked me if he could have permission to go the store to get breakfast, and I told him that it was okay. He took his things off the truck and went to the store. Meanwhile, I continued studying for the examination for Captain. I waited for a long time, and noticed that it was getting close to the time we usually drill. Drill period was everyday at 10 o'clock or 1 o'clock, and this was one of our inside drills. No one came to get me, so I went downstairs to see about breakfast. To my astonishment, they had cooked and eaten without me. I was furious. I was Commanding Officer of the unit, and had given permission for someone to go to the store to get food for the company, and I was excluded from breakfast along with three other blacks. I immediately approached the Captain, and two strokes were pulled which was designed to bring everybody to attention on the apparatus floor. Before I could address them, the truck company got a run. As we cleared the apparatus floor, the firefighters were circled around the Captain telling him their story. Their story had no bearing on what I was going to do, I knew what I was going to do. When I returned, I had the Captain to line them up again, and I made it clear that while we were in that house, we would work together and eat together, or we would all eat separately. I told them if everybody was not going to participate or cooperate, they were not to light the stove. There would be no more cooking. Of course, my directive was an unpopular one, but I was not going to be a part of any form of discrimination.

A firefighter, who was on leave during the breakfast incident, heard about it and decided he would show me minimum respect. He avoided coming face-to-face with me in the mornings to avoid speaking. One morning when he avoided me, I ignored him since it appeared he was going around the other side of the truck. I gave him the benefit of the doubt that he did not do it on purpose. But the second time it was clear that it was on purpose, and I went looking for him. The Sergeant on duty suggested that I let it go because he would only do it again. The Sergeant was right; about a week later, he tried the same thing, and I had my opportunity to deal with him right in the middle of the apparatus floor. It was said that I was supposed to take people to the side to talk to them, but I felt since he was disrespectful in the middle of the apparatus floor; I should deal with him in the same place. I was truly upset, and I gave him a piece of my mind. This guy was a cook. When I finished telling him how it as going to be, relative to meals, he became angry. I told him that the reason he was no longer preparing meals was because of the incident that had occurred about the breakfast. After I explained to him the logic of the incident, ironically, he understood. That firefighter and I became very good friends after that.

A few other minor incidents occurred in that house, but it was okay, because I took care of it, when I learned about it. Firefighters are very unique people, they may have their differences or even be prejudice, but on the fireground, they were one. They care about one another, black or white. White firefighters would give mouth-to-mouth resuscitation to black people. They were united on scenes of emergencies, and performed their duties in an exemplary manner, even under adverse conditions. They worked together.

Although Engine 30 responded a lot, there was still time for fun and games. One morning at three o'clock, while everyone was in bed, a fire-cracker was thrown in the bunkroom. At that time, I did not know that the Captain was in on it, we slept in the same bunkroom. When the firecracker went off, I got up and the Captain asked me if they were disturbing me, and I told him yes. I pulled two strokes to form a company lineup. We started taking hoses off the apparatus, and we changed them all night. That problem was permanently solved. The firecracker scared a rookie firefighter half to death, because he had no idea what was happening. His bed was turned over and his mattress was on the floor.

Early on I appreciated the Captain giving me the chance to prepare for the examination, and I did quite well. Without being able to invest the time, I would not have been as successful. One thing about taking a promotional examination, you must put in a lot of time studying. The amount of material that you had to remember was incredible. At one time, I served on the Promotion Board, and we wrote at least 400 to 500 questions just to get 300 that the board felt would be used for testing purposes. Throughout the years, it was known that studying for an examination meant studying both day and night. I was in a study group for the first couple examinations. There was always one in the group who would never bring anything to the table. He just sat around and absorbed what others had to offer. The group eventually discarded him. Officer Kitt and I worked together for the examination steadily, anywhere we could, on and off duty. We felt that it was essential for us to get promoted because it had become known that efforts were made to try and change the prerequisite for being selected for promotion while assigned to Engine 30.

Chapter 9

Promotion to the Rank of Captain

Having attained the rank of Captain, I was transferred from Truck 16 to Truck 17. This firehouse was referred to as "The Soul Train," because the officers stressed training in a professional manner, resulting in an elite company. Many black firefighters in the fire department wanted to be reassigned to this firehouse, but their requests were denied, because there was no vacancy. Prior to my arrival, a firefighter said that they were called the "Dirty Dozen, because the community had access to the bathroom. A person used the facility and reported to the Battalion Chief that the bathroom was filthy dirty. When this word spread throughout the Fire Department, the firehouse was referred to as the "Dirty Dozen," because 12 firemen were assigned to this station. I said, that would change.

A month later, we were scheduled for an inspection. We had inspections, inspections, inspections. All day long, we were cleaning until inspections were over. The men went underneath the firetruck and cleaned it until it was squeaky clean.

I divided the house in three quarters, numbers 1, 2, and 3 platoons. Each had a job. The first platoon was assigned the left side of the house; the second platoon was assigned the right side of the house; and the third platoon was assigned the apparatus. I developed a plan to paint the walls rather than wash them. In the beginning, there was some opposition from some union members stating it was against union regulations and that painting would interfere with contract painters getting jobs. A few said that they didn't want to wash walls, and others said they did not want to paint. I told them that this was different; they lived in that house, and it was just like their being at home; if they didn't want to paint, then they would wash. The officers decided to obtain plenty of spic and span. I told them washing the walls would not look as good as painting. After some deliberating among themselves, a Lieutenant on another shift came to see me and said, "Captain get the paint." I called the Property Section, and requisitioned 150 gallons of paint, some rollers, etc. I procured everything they needed to do the job. The drivers literally took the truck apart, cleaned all of the tools and appliance, scraped it down, and shellacked it. The truck looked like it was brand new, so did Engine 30. Everyone was excited about our achievement.

The Battalion Chief was Calvin Watson. He didn't say anything and he was not a hard officer. He was the one who was treated so badly in being given the silent treatment early in his career. He knew what the men had done in taking all of the stuff off the walls, etc. They did a beautiful job. The apparatus was sparkling. I was off duty the day of the inspection, but I came in purposely to find out the results. As Captain, I felt that I should be there, during the company's annual inspection.

The inspection officer gave us excellent on the house, and very good on the apparatus. I didn't take the results lightly, and I wrote a letter to the Chief about the rating. When the rating officer received the rating from the Fire Chief, he was somewhat dismayed, because there had never been a Captain to write a letter to a Fire Chief opposing a fire inspection rating. The Deputy Chief responded with his reasons, which was not commensurate with my views. I took the response and placed it on the bulletin board so the men could see it.

I initiated additional training. We were supposed to train everyday at the firestation for 2 hours. The Captain would have a systematic way of training the members on certain aspects of the operation. For example, Tuesday was for air masks, self-contained breathing apparatus. They would put them on the truck, lay them down, and each man would put them on. Wednesday was always outside drills. We had to go outside, stretch hose everywhere, and throw water everywhere.

The Training Academy had a different type of training. They trained for flammable liquids. There was a pit that we would put gasoline in and set it on fire. They showed us how to use CO_2, light water, to put it out. So, there was a method of ensuring that the firefighters knew what was expected of them.

There was another mechanism used to identify what was to take place in a building in our district. We would go to a house, apartment, or hotel and pre-plan. We would say that this building was so many feet high, wide, identify what streets were on the sides, the type of structure it was, how many people were in it, the companies that would respond, which company was due first, second, etc. That building would be surrounded all the way. If we couldn't get to the rear by the alley, then we'd have to drag the hose around. I didn't want the rear to not be covered, because they couldn't get the truck in there. I wanted to go back there and see some hose laid out. I didn't care where it had to come from, I just wanted the hose on the ground. That was pre-planning. If we were due second, we were also supposed to know what was happening. But it was the job of the first due engine and truck company to know their local and box alarm district. The Captain would get the information book, and ask questions about how would you get into the rear of a building at a specific address. They had to tell him. They would write the streets, blocks, etc. on the board. The Captain would critique what they wrote and they would discuss it so everyone on duty would know what to do, in the case of an emergency in that location.

The Captain of the engine company was the Property Officer. Every bolt, nail, etc. were supposed to be on the property card. Every year we had an annual report. The Battalion Chief would come in and look at the desk journal to see if the men were signing in and out. He would go and see where the hose would be stretched out, go in the kitchen and open the cabinet drawers, etc. The Captain was responsible for that. When you had annual inspection, that house was supposed to be ready.

There was a lot that had to be done in the firestations, but we had to have people who were committed in getting the job done. You couldn't tell someone that you were mad at the Mayor or the Fire Chief, saying that they don't care about certain things. The Captain is supposed to be the Captain, and if he hears that kind of talk, he's supposed to take action. But, the Captain could be a member of the union. That's a whole new thing to be considered. Incidentally, I remember the behavior of a Captain at Engine 8. He displayed a sign on his personal vehicle that read, "The Fire Department is not only a job, it's a joke too."

I was at the "Soul Train" for about 5 years longer than anyone else. The Fire Chief called me one day and asked me how long had I been there, and I told him I didn't know. He said he was going to send me to Engine 7 for one year. Engine 7 was less active than most units in the department. It was one of those companies that didn't get many runs until after midnight. However, we had a 10-story building at 1001 M Street that we responded to every night. The occupants used to take trashcans and empty them right at the elevator doors and set the trash on fire. After extinguishing the fire we had to go up eight steps, to cut the water off at the standpipe connection, and proceed to work the fire down. I was told that was the reason they stopped building 10-story buildings for people in public housing. They started building duplexes to keep better track of who was tearing them up. That was a beautiful building. There was another building located at 111-7th Street, a public housing building; and the condition of it was the same as 1001 7th Street S.E. They were both beautiful buildings. In fact, they had their own little clinic where people were able to conveniently get their needed medical supplies. The building became uninhabitable in 10 years.

Chapter 10

Assignment to the Rescue Squad

I was transferred from Truck 17 to Engine 7. The squad personnel considered themselves as an elite type, because they could operate as required in areas that firemen could not. They had special tools that were designed to extricate victims from wrecked or damaged vehicles, enter houses or commercial buildings, and cut through all types of locks and metal entranceways. The Rescue Squad was accompanied by fire engine personnel at all times. The bulk of our calls/runs were to two apartment buildings for fires. The units responding were Engine 18, Truck 7 and Engine 25, Truck 15, with the squad unit assisting where needed.

My duty as a Captain was commanding Rescue Squad 2 and Engine 4, which operated out of a single house. I was in command of all units, until the Battalion Chief arrived. I should have been given the opportunity to participate in this squad when I was at the rank of Sergeant.

The Rescue Squad personnel had to wear a special type face mask, namely a McCoy Mask. It was difficult to operate within itself. However, the Scott Mask was more of a self-contained breathing apparatus, with a pack alarm attached. It was the mask of choice. Its functioning was good for 1 hour. Also, it had an audible warning system when air reached below 300 pounds per square inch. The rescue members would immediately retreat from engagement, and replace their air tank.

There was a particular incident that I was involved with directly. I had to use a one and a quarter inch line, having 150-pound square inch pressure at a house fire located on Mount Pleasant Street, N.W. The temperature was COLD, COLD, COLD. As I sprayed the water onto the fire, the mist fell down onto our running gear, causing a freezing condition that made us look like frozen statues. The Rescue Squad was put out of service, and we were taken to the Washington Hospital Center to be warmed up physically and evaluated for hypothermal. After our release from the hospital, we were permitted to leave for the remainder of the day to recuperate. I returned to duty, and continued to perform responsibly as the Captain, of the Rescue Squad. I also acted as a Battalion Fire Chief, periodically, when directed, in various locations.

There was a firehouse under construction in the North West Section of the city. Soon it was completed and referred to as firehouse Engine 4. Rescue Squad 2 moved in as well, for which I was the Captain.

One day I was called by the Deputy Battalion Fire Chief of operations, who informed me that a vacancy for Battalion Chief was eminent. I was at the top of the eligibility list for consideration and my name was being submitted to the Department Fire Chief. Subsequently, I was informed by the Deputy Battalion Chief that I had been selected by the Fire Chief to fill a position of Battalion Fire Chief, and that my official orders would be forthcoming.

Chapter 11

Promotion to Battalion Fire Chief

My primary duty as a Battalion Fire Chief in the District of Columbia was to operate as the leader at all times, with the companies for which I was responsible. Obviously, the responsibility related to the personnel, as well as the property located in my respective Battalion/District.

The specific responsibilities of my position were to make daily visits to management staff; e.g., hospitals, restaurants, apartment buildings of all sizes and hotels, as well as all commercial buildings having specific usages; and submit a daily report to the Deputy Fire Chief of Operations for his/her review. They attached an endorsement if such was warranted, prior to the report being forwarded to the Chief of the Fire Department. As Battalion Chief, I was required to review any incidents that required the Rescue Squad service, or fire engine activity, to ensure that everything was done in a professional manner, after reviewing the Captain's report.

The Battalion Chief rank was probably the most rewarding for me. Maybe it was because 1 was Captain for 6 years. Many times during the 6-year period, I was detailed to all of the Battalion Chiefs' headquarters when a vacancy occurred. I spent much of those years as a Captain in the 8th Battalion, which is located in the far east quadrant of the city at 49th and East Capital Street. The experience I gained as a truck company officer there gave me valuable knowledge as to what should be expected from companies responding to fire locations. I was well rounded with emergency experience because I was assigned to engine companies, truck companies, rescue squads, and of course, the ambulance service.

I was the Commanding Officer, promoted to Battalion Chief and assigned to the second battalion, which was the largest battalion in the city.

It consisted of Engine 1, Truck 2, Ambulance 1, Truck 5, Engine 5, Engine 23, Engine 9, Truck 9, and Ambulance 3. As a Battalion Commander, I visited each company daily. It was not to see if they were doing something wrong, but to see if they were doing something right. I sat in on drills and inspected apparatus, quarters, and men. Company officers knew what I expected from them, in quarters and on the fireground. When I arrived on the fireground, every position had to be covered. Management on the fireground started at the rank of Sergeant. You learned what the Battalion Chief expected of you. As a Battalion Chief, there was no doubt that all of my company officers knew that all petty situations had to be handled at the lowest possible level. The company officers had plenty of time to mold their companies to perform effectively, both in quarters and on the fireground. Sometimes I sat in on drills and asked questions about the operation. Many times I simply observed the drills for the entire 2 hours to see just how effective the officers were in conducting the drills, not just what the subordinates knew about the subject matter. Everyone knew what the drill was about because it was in the training manual as to what drill would be performed each day. I drilled so much during my period that I knew what each drill was for each day and didn't have to refer to the book. If for some reason there was any doubt, the book was there.

My Battalion always gave a lot and I applauded their efforts because they believed in training. If you train your men/women, they would know what to do on the fireground, because it becomes a part of you. The drill periods were a priority for the department in the Fire Fighting Division. One might ask what kind of drills they performed. They performed drills from tying 2-1/2 hose lines on an aerial ladder to rescuing a person using life lines from the roof, or other areas beyond the reach of the longest available ladder. Drills consisted of 8 hours per week. That's a lot of drilling. The drills were approximately 416 hours a year in addition to the time of drilling at the Training Division. In quarters, we drilled a lot on our local, first, and second alarm districts. Technicians and wagon drivers knew their districts like the back of their hands. If for some reason they missed a turn, they would never live it down.

Chapter 12

Promotion to Deputy Fire Chief, Assignment to Training Academy, and Interim Assistant Fire Chief of Services

In 1980, I was promoted to Deputy Fire Chief, after having been at the Training Academy for approximately 2 years. I was summoned to report to the Office of the Fire Chief. He asked me to take a position as Interim Assistant Fire Chief of Services. There had been some shake-ups in personnel by the Chief. He felt that I had some expertise, knowledge, and commitment, to help him get a job done. At that time, the agency was in such disarray.

When I sat down to discuss the Chief's offer, he made it clear to me that the Assistant Fire Chief of Operations was contemplating retiring. He further stated that if he retired, I would perform as Services Chief, and respond to fires as the Assistant Chief of Operations. Also, he said, at times I may have to fill in for him, when he would not be available. I told the Chief that I didn't have a problem with this arrangement, and that I would do the best job possible under the circumstances. We set out to identify a mode of operation that would be of great value to the Fire Department.

On the morning of January 13, 1982, we responded to a third alarm at 14th and Upshur Street, N.W., D.C. It was a very cold morning; icicles were hanging off the clothing of the firefighters, as they threw water on the building to knock the fire down. When I arrived on the scene, the Deputy Fire Chief was in charge. I immediately assigned him to the rear of the building. Because of the frigid weather, I felt that it was necessary to get some relief for the firefighters who were battling the blaze. I had the Communications Division to contact Metro to make sure that they could send a bus to the location to give the firefighters some relief. Shortly thereafter, the Fire Chief arrived on the scene, and I gave him an overview of the emergency condition. We walked around the building to see what progress had been made. It appeared that the firefighters, Sergeants, Lieutenants, etc. were simply to locate, surround, and knock the fire down. As cold as it was that morning, it didn't appear that any of the firefighters had a problem in carrying out that mission. I was proud of them because I too was in a position to have water sprayed from heavy-duty devices on my coat and helmet, which could have frozen. However, I stood pat in my position as Assistant Fire Chief of Operations. The Assistant Fire Chief of Operations is responsible for ensuring that everyone on the fireground is performing at the peak of their ability. I must say that regardless of what anyone says about the firefighters of the District of Columbia, I believe that they are head and shoulders above any firefighting force in this country.

Shortly after, the Fire Chief and I conversed about the fireground and what had been done, I reported to the office, cleaned up, and prepared for the day's work. Working in positions of Fire Chief and Assistant Fire Chief of Services meant there was a lot of work that had to be done. I had to remain at my desk, glued to it all day to get it done. As we settled down and started getting endorsements and approvals of requests, we had a variety of things to get done. At the same time, the snow was heavy, and it was very cold outside. It was one of the worst times that I had seen since being in the Fire Department. My office was located at 614 H Street, N.W., D.C. As I listened to my fire department radio, I was made aware that a fire existed one block from my office, and it had turned into a second alarm. Under most conditions, I would have responded to the second alarm. The Deputy Fire Chief of the Fire Fighting Division was on the scene. He was a very capable man, so I saw no need to respond to that location. Shortly thereafter, I saw the white smoke billowing above the rooftop. That was my signal that they were getting to the fire. After that, I heard the Deputy Fire Chief notify communications that they had the fire knocked down. When I heard that transmission, I felt relieved because I knew they were taking care of business. So, that day wasn't too much different than any other day.

Subsequently, I received a call from my wife Uvaghn who asked if I had heard about the airplane that hit the 14th Street Bridge. At that time, I had no knowledge of it. I immediately contacted the Communications Division to see if that was true. I was told that they had received a message from someone on a walkie-talkie or CB that a plane had flown into the 14th Street Bridge. It was later confirmed that a 737 plane identified as Florida 90 had crashed on the bridge. At that point, I felt it was highly necessary for me to contact the Fire Chief, and make him aware of the existing emergency. I could not locate him, so, I decided that I would go to the airplane crash site. Prior to my arrival, I was made aware that a train had crashed in the Metro tunnel at the Pennsylvania Avenue, N.W., D.C. station. In the process of determining which emergency to respond to, I opted to go to the airplane crash site. It was tough trying to get from my office to the 14th Street Bridge. Anybody who was around at that time could attest to the fact that 4:00 o'clock in the afternoon was in itself a serious challenge without the snow. It was one of the worst things that could have happened to our city. Upon arrival, I went to the Deputy Chief, who had placed a squad wagon on the bridge as our communications center. I went down near the water. I couldn't see any value that the squad wagon would be to me that far away, and directed the Deputy Chief to have it placed closer to my location. The Deputy Chief informed me that it was where he wanted it. Without any hesitation, I asked the Deputy Chief if he knew who was in charge of the scene. His reply was that both of us were Deputy Chiefs and as he understood it, he was to remain in charge until a higher-ranking officer responded to the location. I said to him that he had just answered his own question, and that I was the higher-ranking officer on the scene. He seemed reluctant to accept my explanation. I asked him if he would get the squad wagon down, and in place where I wanted it, or did I have to make some adjustments. One of his colleagues standing there whispered in his ear that he had "better get the squad wagon down here because TR is not playing." He was right. He came about 1/16 of an inch from being suspended. I had worked in this job for a long time, and I gave respect to every officer above my rank. There were times that I gave respect when it wasn't due or needed. That's the way I was raised. I was raised to respect people. Of course, the squad wagon driver was directed to bring the squad wagon to the more conducive location.

We then established one of the squad wagons as the communications center to transmit messages from the Communications Division to the Fire Chief. In the tragedy of the Air Florida crash, people were being rescued from the water; and some were taken to hospitals, and others to a makeshift morgue. In a situation of that kind, none of us had ever seen an airplane crash. We had trained a lot to the point of what we would do with large volumes of diesel or aviation fuel being on fire. That's the biggest problem that firefighters think about during an air crash emergency, large volumes of aviation fuel burning. With this situation, we didn't have that because there was no fire. Part of the aircraft was in the river. So, there was a lot of salvage work, and rescue operations placed in service. Every Fire Chief in the Metro area, and maybe further away, was on the scene that day. It was amazing that I only knew a few of them because of all the meetings that I had been to. I recognized the Chief from Arlington, Va., Chief from Fairfax County, Va., and Chief from Prince George's County, Md. I had seen the others, but didn't know who they were. In my mind, I knew that something had to be done so that the Fire Chiefs in Washington Metro area would at least know each other when they see them.

Shortly afterwards, it was nearing night fall, so I decided to leave the Deputy Fire Chief in charge of the airplane crash area, and I immediately responded to the Metro train crash at 12th and Pennsylvania Avenue, N.W., D.C. When I arrived at the scene, I could see that the train had indeed jumped the tracks, and the rescue operation was in full force. The D.C. Fire Department personnel act responsibly when the chips are down; they take care of business. The Emergency Ambulance Service played a vital part in that rescue operation, because they were transporting people to various hospitals in the city as quickly as they could.

When things had settled down somewhat, and the rescue operation had subsided, I felt that it was time for me to go home. I had been on the scene since 3:00 that morning, and it was now 6:00 in the evening. That would give me about 16 hours of continuously doing my job. I went home, got myself cleaned up, had dinner, and went to bed. The next morning being Saturday, I got up and headed for the office, because I didn't know where the Fire Chief was at the time. When I got there, he was there in the office. He and I talked about a few things pertaining to the operations that had been completed, and we decided to revisit the airplane crash site. I had never seen anything like that before in my life. That morning there were people still floating around in that murky, cold water, because the rescuers were unable to get them all out. When you are in an emergency operation such as the Fire Department, you see these kinds of deaths taking place all of the time. You get in there and do whatever you can to save a life if possible, sometimes under extreme adverse conditions. Because it was Saturday, there wasn't much we could do. The salvage operation was in full swing, so I went back home and took it easy the rest of the day.

I went back to work on Monday morning, and the Fire Chief was on leave, so I was performing duties as the Fire Chief, Assistant Fire Chief of Operations, and Assistant Fire Chief of Services. Of course, I had the Administrator, there with me, who worked for 3 or 4 Fire Chiefs over the years. He was quite knowledgeable about the operation of the Fire Department. I used his expertise and knowledge in whatever areas so I could keep things running smoothly.

I received a call from the Mayor's office, and they wanted to speak to the Fire Chief, who was still on annual leave. As I understand it, the City Administrator wanted him to cancel his leave, and come to his office. I was able to contact him, and I relayed the message that he wanted him to come down to his office, because he wanted to talk to him. The Chief's reply was that he was on leave, and that he would not be going to see the Administrator. As the day progressed, not much happened. The next day, the Chief called me and asked if I would do him a favor. I told him sure. He asked me to tell the Administrator that his papers would be on his desk tomorrow morning, he was retiring. I used to call him "Big Rich." I said to him, "'Big Rich,' you can't do that." I asked him to come into the office so we could sit down and strategize. I told him there was a lot of work that had to be done, and we could do it together. We talked about the leave situation, and I suggested that he not retire, but deal with the issues, and not run from them. He said no, he was retiring, and asked that I convey that to the Administrator. This was about 10:00 or 11:00 a.m. Later that day, he called and asked me if I had called the Administrator to deliver his message, and I told him that I had, but he was not in his office at that time. He told me that he would call back. I had not tried to get the Administrator, because I was hoping that I could convince the Chief to come back to work so we could take care of business. It didn't happen. The third time he called, I decided that it was time for me to deliver his message. I called the City Administrator and relayed the message from the Chief that he was retiring.

The City Administrator said the Chief was out of his mind, and that he could not do that. I told the City Administration that I conveyed the message as instructed. He asked me to contact the Fire Chief again, and have him call his office. I explained to him that the Fire Chief had turned his pager off, and that he was in the process of forwarding his retirement papers to the Mayor for his approval. Obviously, the Mayor had gotten the message from the Administrator; within 15 minutes, the Mayor called me to inquire as to what was going on. I told him that the Chief felt that the Administrator was leaning on him too hard; also that he was being disrespected and felt that he could not work under those conditions. What he meant by that, he never discussed with me. The Mayor told me to have the Chief call him. He said that we didn't need to work that way, because we can take care of anything. I told the Mayor that I would deliver the message. Later that day, the Chief called me, and I conveyed the Mayors message. Also, I said, "'Big Rich,' call and talk to the Mayor. Tell him what your problem is concerning the City Administrator." He said he didn't know of anything that the Mayor could say to convince him that he and the Administrator were not together in an effort to fire him. I asked him what had he done to be fired. He simply said, "I am retiring."

Within a few days after my discussion with the Chief, I received a telephone call from the Mayor, between 5 and 6 o'clock in the evening. He asked me to come down to his office. I put on my eight-button box (uniform), and complied; he and the Administrator were in the office. The Mayor said to me that it seemed as though the Chief is going to retire on us. I said, "Yes Mr. Mayor, it seems that way." The Mayor then said, "We are going to need a Fire Chief." He asked me if I thought that I could handle it. My response was, "Yes, Mr. Mayor, I can handle it." We had a question and answer session in the office for well over an hour concerning a variety of issues; e.g., morale, policies, promotion, apparatus, overtime, budget, as well as subjects of a more mundane nature.

The Mayor said, "Ok, we'll make you Acting Fire Chief today." The Administrator intervened by saying that I was currently the Deputy Chief at the Training Academy. The Mayor directed him to make me a permanent

Assistant Fire Chief and an Acting Fire Chief, the same day. The Mayor said he would sign the papers.

The Mayor told me that I would have to select both Assistant Fire Chiefs of Operations and Services. This was my call. The Mayor said that whomever I selected I would have to work with them. I told him I didn't have a problem with it. I went back to my office and thought for a few minutes, because I already knew what I was going to do. The next day I called two Deputy Chiefs in my office and told them that the Mayor had given me an order to select an Assistant Fire Chief of Operations and an Assistant Fire Chief of Services. I told them that they were the ones I was going to select for those positions. They were the most qualified on the register. Things went fairly well.

After all pertinent paperwork became official, the new team encountered many problems, because the Fire Department was in shambles. Some of the problems were with the Assistant Fire Chief of Operations and the Ambulance Service, inoperative apparatus due to maintenance, morale, and so on. My work was cut out for me, and I knew that my management style was not going to be popular; but it did not matter.

I looked at the ratio of blacks and whites in the department. It was ridiculous to say the least. There were only a few blacks in a few places. Out of approximately 35 Battalion Chiefs, 5 were black. Disparity existed in all ranks. I decided to make promotions in order to create some equality within the department. As it was, anyone could readily see that something was wrong with the racial and promotional makeup. It could have been identified as discrimination or disparity in the promotional system, or any other reasons. In the process of an attempt at leveling the playing field, with respect to a form of transparency, I was met with court action. If I promoted blacks, the whites would sue, if I promoted whites, the blacks would sue, and I would have to endure long drawn out court attendance time.

The Mayor had directed me to implement an affirmative action plan for the department. The plan had a 5-year turnaround time to have a percentage of blacks in the department. I believed it could have been done; however, the union was an obstacle. I dealt with the union frequently, because I was a member of the contract negotiation team for the department. One of the issues was the Assistant Fire Chief would submit to the Fire

Chief, nine names, plus the number of promotions to be made. The Fire Chief would select from that group of qualified individuals. When I promoted some blacks from that number, I was criticized by the union immediately, even though my actions were governed by the contract. The union maintained that criticism was not their intent. A series of meetings were held in the City Administrator's Office to negate the promotions I made. I was directed by the Administrator to sign an agreement to promote whites, who felt they were not selected due to reverse discrimination. I was vigorously opposed to this. That was one of the first issues that I encountered with the union. From that point on, there were no tranquil agreements between the Fire Chief and the union. Nonetheless, I felt that there was a place and a need for organized labor in the department, even though I was under attack daily with the union's recommendations to successfully bring some promotional transparency to the race issue within the department. I strongly felt that the Fire Department and the City Government were giving up too much to the union. For example, senior managers or supervisors belong to labor unions. There was no span of control in this. Any Sergeant, Lieutenant, or Captain belonging to the union could lodge a complaint against a Captain who was a senior manager that said or did something they did not like. The union then dealt with the Captain. It made no difference because the Captain paid dues as well. This was definitely counter-productive. My Assistant Deputy and Battalion Chiefs were all members of the union. This made it difficult to manage the department. On several occasions, I complained to the Mayor about it. I heard the Mayor tell the Chief negotiator to get the Captains out of the union, but it was never done. Why it was not done was beyond me. The Police Department did not have high-ranking officials belonging to the union, only Sergeants. What was so different between police and fire? Why was there double standards in the nation's capitol?

The Fire Department needed strong leadership during this transitional period. I was not afraid to address issues that were contrary to policy, rules, and regulations. They were strictly enforced.

I had been in the department for over 30 years with an unblemished record; no one could find anything derogatory in it. I made rescues, advanced through the ranks, and was even injured in the line of duty. That did not mean anything. In a way, people act just like some vicious animals.

They will do anything against another person if they thought they would get something out of it. The only thing that they could accomplish would be self-satisfaction. This is what they did to me, and I think their main reason was for me to retire. Their opinion of me was based on totally false information that they received from other individuals, both internal and external. People with personal vendettas called the media and told them petty things, no matter how large or small. They kept things going relative to my managing the Fire Department. The media followed ambulances around to take pictures to the point where the crew could not get into places before the camera crew arrived. Nonetheless, I weathered the storm. I was frequently asked by many how I did it. They said I could take more whippings than a Georgia mule. That was a lot, however my agenda was nothing more than to create positive changes in the Fire Department. The Fire Department created this situation, and I had to pay for them. It is a proven fact that anybody who strives to do good in a racist organization will have to pay. They are not going to allow you to do good when there is a black man involved, unless he pays. I do not feel too bad; I am no different than anybody else. Some people will pay a much greater price for bringing change and equality like the great Dr. Martin Luther King, Jr., and many others. He paid with his life. No, I was not like Dr. King, but the principles were the same in trying to make change for blacks in the D.C. Fire Department. Never in the history of the department were so many blacks elevated to management positions. Let me make it clear, I did not promote them primarily because of their race; I did it because they were qualified, and possessed the knowledge to manage their departments. I promoted whites also. I heard someone say, if you are going to get a decent tune from a piano, you must have both black and white keys, and I understood that.

From the beginning of my appointment as Interim Acting Fire Chief, the system was designed to create unrest for me; and I was attacked on many occasions. I remember a young firefighter was disrespectful to me in an official company lineup. I shook everyone's hand, but when I extended my hand to him, he refused to shake it. I took him in the back for a private conversation, all to find out that he was upset with me because he thought I kept his brother from being hired by the Fire Department. I did not know his brother, nor did I know where he was on the hiring list. The D.C. Office of Personnel did the hiring; I had nothing to do with it. I would however, swear new firefighters in after they were hired. This firefighter was surprised to discover that the Fire Department did not have its own personnel section. He felt bad, and later he shook my hand.

Having encountered some of the most challenging problems in my appointment as the Acting Fire Chief, I considered retiring. However, I was offered the full position as Fire Chief in the District of Columbia by the Mayor. I gave it some thought and discussed it with my wife Uvaghn. She said, "You know you have never run from a challenge." That was my cue to accept the position. Also, I strongly believed that with my overall years of experience in the Fire Department, I could make a difference in the department's modus operandi.

Chapter 13

Assuming the Fire Chief Position

In anticipation of being sworn in as Chief of the District of Columbia Fire Department, it was a time in my life when I felt like a real honored celebrity. We had planned for this special occasion to take place on June 2, 1983; and in doing so, invitations were sent out to friends and relatives in cities and towns, both near and far. Because of the cost factor, my aid suggested that the bulk mail system in the Fire Department be used at 17 cents per unit. I agreed; plus I wanted to be frugal.

Within a week or so of the event, my wife Uvaghn called some of the invited guests to see if they had received their invitation to my swearing in ceremony; they had not. Upon hearing this, an all out points effort was started to determine what had possibly gone wrong. It was much to my chagrin that I learned the mail had been returned for insufficient postage. When we learned of this, we called many of the invitees by telephone to extend an invitation; and the response was simply laudable. The program alignment was carried out to perfection.

The Mayor put out a news release saying that he would officiate at the swearing in ceremony of my being selected as the 20th Fire Chief of the District of Columbia Fire Department. The ceremony would take place at the Office of Personnel Management Auditorium, 1900 E. Street, N.W., Washington, D.C. The news release went on to say that the Mayor, other city officials, and members of the community would join to celebrate this meaningful occasion. It also referred to my being a highly respected official, culminating 30 years of dedicated service to the District of Columbia. This release was relayed to have media coverage at the swearing in ceremony.

On June 2, 1983, at 9:30 a.m., my wife Uvaghn and I positioned our-selves on the side of the building on Virginia Avenue, between 19th and 20th Street, N.W. A Captain opened the door for us to enter, and gave my wife Uvaghn flowers. She and I walked under the aerial ladder arch that had been erected, and preceded to the line of firemen that we would be marching through. Two flag Honor Guards positioned themselves in the center, and escorted my wife and I through the line of firemen. When the two flag Honor Guards reached the last fireman in line, they stopped, and stepped to the side. Immediately in front of us, there were two other Honor Guards. They led my wife Uvaghn and I to the auditorium. One of the Honor Guards took my wife by the arm, and led her to the front row center seat in the auditorium. The Honor Guard returned to the auditorium en-trance and stood beside the other Honor Guard. The Honor Guards, along with the Flag Guards, took up position at the entrance of the aisle.

An announcement was made from the podium, by The Public Affairs Director saying, "Ladies and Gentlemen, may I present the Fire Chief Designate of the District of Columbia." I began walking toward the front stage, preceded by the flag and Honor Guards. The flag Honor Guards stopped at the center front of the stage, and the two Honor Guards escorted me to my designated seat on the stage. The two Honor Guards returned to a position of attention beside the Flag Honor Guards. An announcement was made from the podium, "Ladies and Gentlemen, the National Anthem."

As I stood and peered out into the auditorium, filled to capacity, I cannot over emphasize how elated I was in seeing so many of my friends, relatives, and well wishers, who had come to witness my being sworn in as the Chief of the Fire Department, for the greatest city in the world.

My brother James was seated on the dais with me, along with the Mayor of our city, Marion Barry, members of his cabinet, and my top level managers.

My Public Affairs Officer at this point of the proceedings had done a magnificent job; and if the remainder of the program of the day was as grand, it would be like the swing of the golf club to the ball on the green, resulting in a hole in one.

The Public Affairs Officer announced, "Ladies and Gentlemen, wel-come to the installation of this cities 20th Fire Chief Designate, Chief Theodore R. Coleman. Also, I would like to introduce to you the master of ceremony for this morning's program, the Editorial Director of WDDM-TV, a certified Professional Medical Technician, and a very, very, good friend of the Fire Service; Mr. Rick Adams."

"Thank you, may I ask you to rise again for the invocation, which will be presented by Elder James G. Coleman of the United House of Prayer, the Fire Chief Designate's brother.

"Almighty God, our heavenly father, the author of all good, and the giver of all gifts and greatness. Lord we come to say thank you for your saving grace. The grace that has brought us safe this far, the grace that will lead us in. Lord, we give the hardy and sincere thanks for the manifold blessings that thou has bestowed upon us, and for this moment where we have assembled to witness swearing in ceremony of the Fire Chief of the District of Columbia, T.R., one who has dedicated his life service of this city for more than 30 years. We beseech you heavenly father for thy continued blessings upon him, and his family. Give him the wisdom to rule and govern the Fire Department for the citizenry of our city and to ensure the safety of its' citizens. May the invisible presence watch over each of us, and in the light of a new day, may we rise to bless thee for thy shelter and care. Make Satan behave, this is our prayer in Jesus Christ our Lord. Amen."

Emcee Adams:

"Thank you Mr. Mayor. Chief Coleman is known throughout the Fire Department for his 30 years of service, and to share with you now, some of the highlights of this, is an officer in the Fire Department, Captain Ray Alfred."

Mayor Barry:

"To Chief Coleman, Mrs. Coleman, the entire Coleman family, and to the friends of the D.C. Fire Department, I am so delighted to be part of this ceremony. In April 1982, I had the honor and the authority to designate Theodore R. Coleman to be the Acting Fire Chief of the Fire Department of this city. I gave him some very simple directions. Be responsible to the citizens of the District of Columbia, provide the best pre-hospital emergency medical care possible, and protect them and their property from fires. These words were clear and uncomplicated.

I will not take these words very lightly, nor did Chief Coleman. He has chartered a course for achieving our common goal for the Fire Department, and has already taken great strides in this direction. He and I both know that we have the finest fire service of any city in America already (applause). His philosophy is similar to my philosophy that we will never be satisfied even if we are the best; we want to get better, and we're going to continue to do that.

We talk constantly on the telephone about matters. We see each other, uh, he spent over 30 years in the D.C. Fire Department. He came into the department when it was tough, particularly if you were black, or a woman. And so he has worked in every aspect of the Department on both the operational side and the firefighters side, as well as the services side. His knowledge of the department and the managerial skills he developed during his years as service will be valuable tools in his new assignment.

I know that his true devotion to duty and insatiable drive, and workaholic that he is, will continue with innovation, and the loyalty to which he has served in the past year as Acting Fire Chief.

We have a fine Fire Department, as I said earlier. We could do a lot better, because I just believe in being the very best. If there are to be any class, let's be first class, and we have a first class Fire Department, and a First Class Chief, a first class family and friend of the Chief, and first class men and women who make up the Fire Department.

I am so proud of them, and I am so proud to be the Mayor of the greatest city in the world, to the greatest Fire Department in the world. Thank you" (loud applause).

<u>Biography of Fire Chief Coleman By Captain Ray Alfred</u>

"I have the pleasurable task of introducing the Chief, but before I do, I find it imperative that I preface my remarks with the charge of Congressman Walter Fauntroy to Mayor Marion Barry and the City Council at the Inaugural Breakfast regarding the necessity of using compassion. The person I am about to introduce is synonymous with compassion. He understands that unless our men and women work together this agency cannot be effective in its delivery of service to the community. He expects and demand perfection in the delivery of that service.

Permit me to expound a little on his background. He was born in Danville, Virginia. Shortly thereafter, his family moved to Durham, North Carolina. He was educated in Leighs, South Carolina, where he graduated in 1944. He was honorably discharged from the U.S. Army in 1946. He later joined this quasi-military organization, the District of Columbia Fire Department, in 1953. After his appointment, he immediately established his goals and began working toward accomplishing those goals.

Chief Coleman's related practical experience included a broad scope of boards and committees to identify departmental problem areas and to provide or recommend alternative solutions. These functions were also designed to promote a smooth and efficient operation within the Fire Department.

Since his appointment to this organization, he has been promoted to all ranks up to and including the Fire Chief, and has been assigned, reassigned, and detailed to various other positions in a fire fighting and/or administrative capacity, and, in 1982, was selected by Mayor Barry to be the <u>Acting Fire Chief.</u> As Acting Fire Chief, he made a number of significant accomplishments; i.e., just to name a few, in his compassion for others:

- He recognized a need to honor the 4 million men and women throughout this country who risk their lives for the sake of others. So, last year he organized and this city hosted the 1st

Annual National Firefighter's Parade to honor our nation's firefighters.

- He realizes that the city is undergoing very austere times and to help insure that it survives fiscally, he lobbied for and was successful in securing the 1988 convention of the International Association of Fire Chief's, to be held at our new convention center. This will bring some much needed revenues to the city.

- After the January 13, 1981, metro crash, he saw a need for better coordination in our rescue efforts. So, he ordered the planning and implementation of the first Mass Casualty Metro Drill. This year the drill was successfully implemented.

- Due to Chief Coleman's personal concern, he committed himself by arranging meetings on a regular basis to establish a forum for discussion of intradepartmental exchange of information with area Fire Chiefs, such as, potential hazards and possible disasters involving our metro system. As you can well understand, disaster knows no boundaries.

- Finally, as yet another accomplishment, for the first time in my 20-year career, I am not mistaken for a bus driver, or a policeman, or Navy personnel. Chief Coleman personally designed the shoulder patch that we are now wearing on our uniforms.

In early 1983, Mayor Barry through his wisdom chose this individual to fill the vacant position of Fire Chief, and on March 16, 1983, by resolution 5-63, he was confirmed by the City Council."

MC: "Ladies and gentlemen, may I present to you the Fire Chief for the District of Columbia, Chief Theodore R. Coleman, who will receive the Oath of Office from the Honorable Luke C. Moore, Judge, D.C. Superior Court. Also, we would like to ask his immediate family to come and stand with him while the oath is administrated."

Oath of Office by Honorable Luke C. Moore, Judge, D.C. Superior Court

"Raise your right hand. Are you ready to take the oath of office?

I, (repeat your name), Theodore R. Coleman, do solemnly swear that I will support and defend the Constitution of the United States and preserve the laws of the District of Columbia against all enemies, foreign and domestic; that I will bear true faith and allegiance to the same. That I take this obligation freely without any mental reservations or purposes of evasion, that I will and faithfully discharge the duties of the Office of Chief of the Fire Department of the District of Columbia, of which I am about to enter, so help me God."

After my swearing in by the now deceased Judge Luke C. Moore, I got a deluge of congratulations from attendees. When the hugs and kisses were over, I went to the podium, and addressed the audience. I told them that I understood when you get up to speak in front of a large audience such as this, you're supposed to be at ease, content, no butterflies in your stomach, and that I had studied a five-step formula to prevent this from happening, but it seems not to be working today. (The roar of laughter from the audience was tremendous). Immediately thereafter, I gave my acceptance speech.

Remarks by Fire Chief Theodore R. Coleman

"Mayor Barry, members of the City Council of the District of Columbia, honored guests, officers and members of the department, family friends and well wishers. I am honored by your presence and grateful that our Mayor has bestowed upon me the honor of managing and guiding this great city's fire service.

To be successful, in any major system whether it's a fire system, an army, a football team, or as in any business, requires of its members a dedication of purpose, a spirit of unity and co-operation, and a sensitivity of one worker for another. I accept this responsibility with great enthusiasm, as I dedicate myself and our department to the safety and protection of the citizens of the District of Columbia, as well as the visitors to our great city. Enthusiasm is the prime requisite for success in every area of human endeavor. The dictionary defines enthusiasm as inspiration ardor, fervent zeal, intense desire, feeling, or emotion. It is vital that the enthusiasm that I possess is permeated throughout the department. I might add, enthusiasm links knowledge to purpose and gives it driving force.

I intend to instill a spirit of unity, sensitivity, and cooperation among all members of the department with an intense desire to be worthy of their trust and to exemplify what they are. For their strengths can compensate for my weakness and their wisdom can help minimize my mistakes. If we are to be a potent force in our profession, and the flagship of this nation's fire fleet, not only must we work together, but we must foster and live the essentials of fellowship and brotherhood, with a genuine compassion for each other.

I believe, that both individually and collectively, we possess the perseverance to let our past mistakes create a resurgent commitment to the basic principles of our department. For each of us knows well, that if we are unable to display genuine compassion for each other, we have no future.

Many of us can recall times when we stood magnificently united. In those times no prize was beyond our grasp. But, we cannot rest on our laurels; we cannot afford to drift, we cannot afford to do anything which will cause disruption in our department, nor can we afford to lack boldness to meet present and future commitments. Together in a spirit of individual sacrifice, for our city, our families and ourselves, we must simply do our best.

There is a strong conviction among us as a peace-loving people, a conviction that says that differences in a democratic society are easily resolved if the various sides are willing to meet halfway. I want you to know that I am not opposed to that philosophy in some quarters. But, in the business of fire safety and in the business of protecting the lives and property of our citizens, I draw the line. For this is a point on which there is no bargain. We are not in a game of chance, where there are winners and losers. We are in the business of providing the best possible service to each and every citizen, every day of the year, and on this issue there is no room for compromise. We are cognizant of the fact that we are a service organization and our primary function is to render the most efficient services to our communities. I embrace the concept of doing the best possible job for a good day's pay. I do not feel that the citizens of this city should settle for anything less.

I believe that each member of the department within hearing distance of my voice will attest to the fact that we belong to one of the most rewarding professions in existence. We will continue to work hard together so that the citizens of our communities can realize the benefits of their tax dollars, and visitors to our city can wear the feeling of safety.

I stand pat on my conviction that our Fire Department is destined to remain the No. 1 fire suppression force in the country. In my 30 years of service in the department, I have seen discrimination, disparity treatment, inconsistency in operation, and under some conditions, just plain old disrespect for each other. In my role as Fire Chief, I intend to eradicate these inequities wherever they exist.

It is my hope that when my tenure as Fire Chief has come to an end:

- That we have served this city well, and effected major accomplishments

- We have improved morale, and enhanced organizational pride

- We have torn down barriers that separated those of different races, natural origin, sex and religion; and where there has been mistrust, we have built unity and respect

- We have ensured equal treatment for all

- And that our performance has made you proud.

As members of an emergency organization working sometimes under extreme adverse conditions, no matter how arduous the task, we will never quit until we have accomplished our intended mission.

Thank you."

Rick Adams (Emcee)

Rick Adams: "Ladies and gentlemen, we have an addition to the program, a presentation of a plaque to the Chief, by Mrs. Jane Mackie. Mrs. Mackie will you come forward, and make your presentation now?"

Mrs. Mackie: "Today, I am happy to be here; you have paid your dues, and I am happy that you didn't forget along the way that I am your adopted sister. To the honorable Theodore R. Coleman, congratulations, happiness, and best success always. Thank you."

Chief Coleman: "Thank you very much Jane. I didn't expect this, but it is marvelous. Thank you" (applause).

Rick Adams: "May I ask you to rise and offer the benediction. It will be presented by Dr. William E. Bishop, the Protestant Police, and Fire Chaplain."

Dr. William E. Bishop: "In closing, Chief Coleman the road will rise to meet you, and the wind will be always at your back, may the sun shine warm upon your face; the rain fall soft upon your fields; and until we meet again; may God hold you in the palm of his hands. And now unto God's gracious protection, we commit you and all of us, and the blessings of God Almighty be upon us and remain with us always. Amen."

Rick Adams: "While you are standing, may we have one more round of applause for the 20th Fire Chief of the District of Columbia."

After the inauguration ceremony, my wife Uvaghn and I were transported to the Hilton Hotel ballroom, followed by other family members, my staff, and well over 300 well-wishers who engaged in a festive time mingling while sipping champagne and sampling hors-d'oeuvres. A Police Department Band, called Side by Side, played the music of rhythm.

I was extremely elated to have my mother in attendance. She suffered with a severe condition of crippling rheumatoid arthritis. (She is now deceased).

Following the event at the hotel, my wife and I were transported to our home to relax and freshen up, before going to the next planned event, which was to be a formal affair at the Norwegian Embassy. Approximately the same number attended as at the Hilton Hotel earlier in the day. The music was from a live band. Champagne and hors-d'oeuvres were served. It was a joyous affair, very colorfully done. Again, my hat's off to my Public Affairs Officer, Michael Tippet, for his splendid coordinating of the events for the day.

I would like to share some congratulatory letters sent to me on my appointment as the Fire Chief for the District of Columbia:

Jerry C. Moore, Jr., Councilman: Dear Chief Coleman, I was away from the city on June 2 attending my mother's funeral, and therefore, could not join you and your many friends for your swearing in ceremonies, and the two receptions given in your honor. Congratulations, and keep up the good work.

Maurice T. Turner, Jr., Chief of Police, D.C.: I regret that I will be unable to join you and your colleagues at your swearing in ceremonies as Fire Chief. However, I will be out of the country on June 2. I wish to congratulate you on becoming Chief, and look forward to working with you on mutual concerns and interest. With kindness and best regards.

M.H. (Jim) Estepp: Dear Ted: Congratulations, on your recent appointment as Fire Chief for the District of Columbia. I know you will continue to enhance the Fire Department Operations with your fine qualities and leadership abilities. I am sorry I was unable to attend your swearing in ceremony. Unfortunately, I did not receive the invitation until the afternoon of June 2. Good luck in your future endeavors. If I can be of any assistance to you, please do not hesitate to contact my office.

Alfred A. Saviaa (Acting Chief, Fire and Rescue Dept.) County of Fairfax, Va.: I received the invitation to your swearing in ceremony today, and needless to say, it was impossible for me to attend. I certainly am sorry that I didn't know about this occasion sooner, because I most certainly would have been there. Congratulations on your appointment, you are most deserving, and the best of luck in all your future endeavors with the department. If our department can be of any assistance to your department at any time, please don't hesitate to call on us. Again all the best to you.

Yolonda Smith, MD: Dear Chief Coleman: Please allow me to add my congratulations to those of your many friends and family. When we met last month, I did not realize that you were "acting." Best wishes for a long and successful tenure as Chief. The District is fortunate to have you in this position. I look forward to your joining us for a transport. (Georgetown University Hospital, Director of Transport Division of Neonatology).

Sal Edelsteinn, MD (Director), Mark Smith, MD (Associate Director), Robert Shesser, MD (Associate Director), Shirley Adams, MD (EMS Coordinator). All four Doctors are from George Washington University Medical Center. Dear Chief Coleman: The facility and staff of the George Washington University Medical Center, Division of Emergency Medicine, wish to extend to you our sincere congratulations on your confirmation as Fire Chief of the District of Columbia.

We are appreciative of your support and commitment to the Emergency Ambulance Division, and the EMS at the local and regional levels.

We are looking forward to working with you to achieve the goal of providing the very best in emergency medical care, to the citizens and visitors of the District of Columbia.

Thomas M. Hawkins, Jr.: Dear T.R., congratulations on being sworn in as the Fire Chief of the District of Columbia. The mayor selected an outstanding individual to be the number one person in the department. I can pledge my continued cooperation to foster a good working relationship between two jurisdictions. You can be proud of your efforts to bring Arlington and Washington close together.

I apologize for not attending your swearing in ceremony. I would have attended if it had not been for the postal service mistake. We can fly people to and from the moon, but mail cannot be delivered efficiently across the Potomac River.

Again, best wishes. If there is anything I can do to assist you, please do not hesitate to call. (Fire Chief, Arlington County, Va).

D. Robinson, Inspector: Sincere congratulations intended to express a wish for your continued luck, success, and happiness!

Robert L. Lindeberger, D.C. Fire Department (Ret.): Warmest wishes to you for good luck, great success, and happiness, all the things that someone like you always deserves. Thanks for your imaginative, creative leadership of the District of Columbia Fire Department.

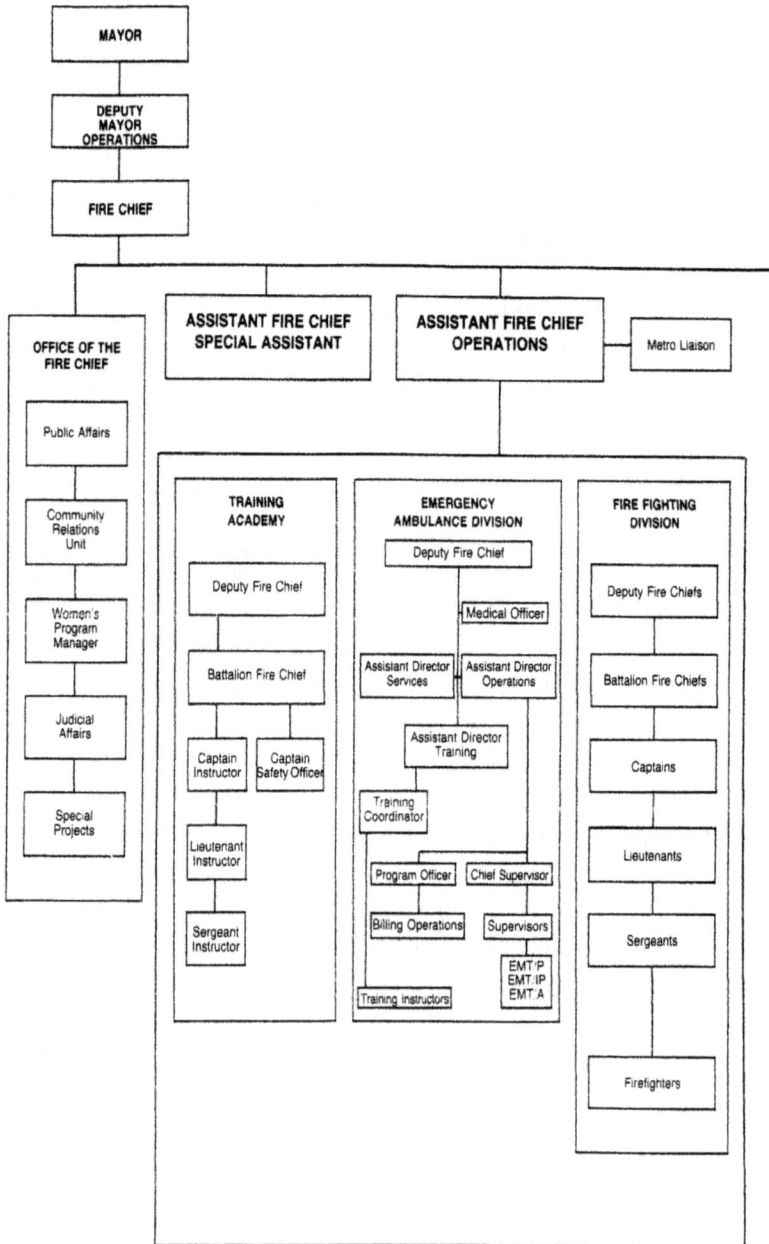

Organization chart.

- MAYOR
  - DEPUTY MAYOR OPERATIONS
    - FIRE CHIEF
      - OFFICE OF THE FIRE CHIEF
        - Public Affairs
        - Community Relations Unit
        - Women's Program Manager
        - Judicial Affairs
        - Special Projects
      - ASSISTANT FIRE CHIEF SPECIAL ASSISTANT
      - ASSISTANT FIRE CHIEF OPERATIONS — Metro Liaison
        - TRAINING ACADEMY
          - Deputy Fire Chief
          - Battalion Fire Chief
          - Captain Instructor / Captain Safety Officer
          - Lieutenant Instructor
          - Sergeant Instructor
        - EMERGENCY AMBULANCE DIVISION
          - Deputy Fire Chief
          - Medical Officer
          - Assistant Director Services / Assistant Director Operations
          - Assistant Director Training
          - Training Coordinator
          - Program Officer / Chief Supervisor
          - Billing Operations / Supervisors
          - EMT/P EMT/IP EMT/A
          - Training Instructors
        - FIRE FIGHTING DIVISION
          - Deputy Fire Chiefs
          - Battalion Fire Chiefs
          - Captains
          - Lieutenants
          - Sergeants
          - Firefighters

90

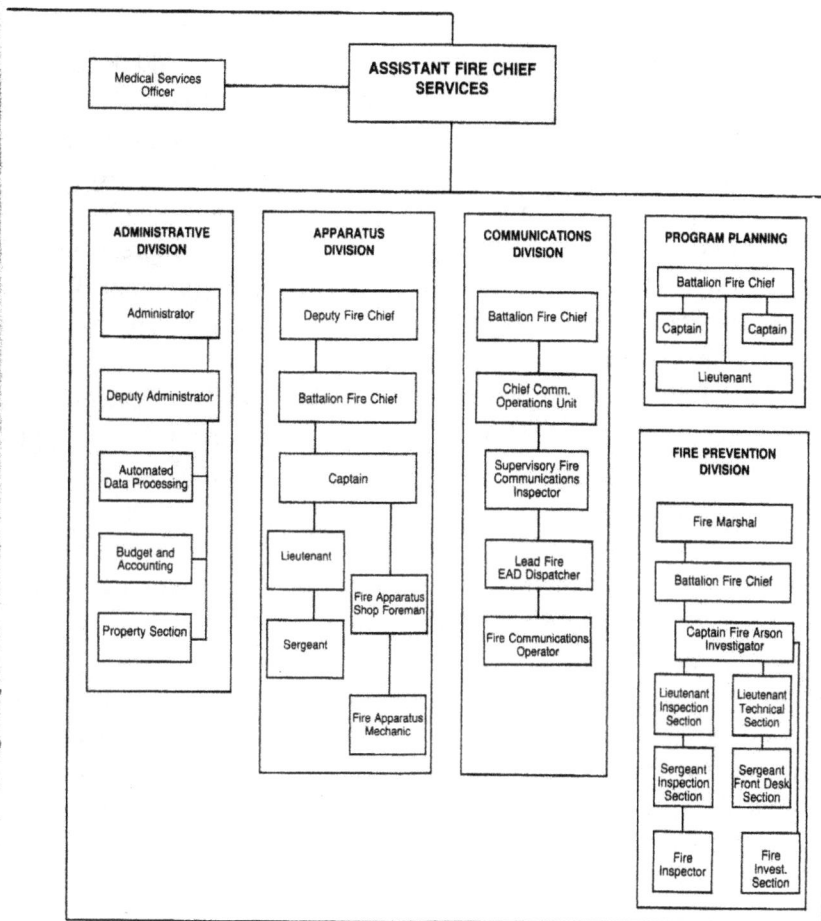

Chapter 14

From the Top Down "Frankly Speaking"

In operating a Fire Department, there is one thing that should be identified as a priority—fire prevention. There is no mystery or substitute for fire prevention. If we prevent fires from occurring, we minimize approximately 90 percent of the budget allocated for fire suppression. The Ambulance and Fire Prevention Division is the cornerstone of any Fire Department. We devote a large amount of time and effort to fire suppression when much of it could be directed towards a fire prevention program. We have not concentrated on fire prevention, to the point of using firefighters to assist through the fire prevention effort.

We have concentrated on fire safety in the schools to teach children how to be protected in their homes; we have taught them how to explain safety tips to their families that they have learned through the efforts displayed by community relations. Community relation organizations are credited for teaching children how to take information learned from the fire department back to their homes. Many times I have heard families talk about what their children say to them about fire prevention: how to develop an escape route, how to know where all of the exits are, how to get out of a burning building, how not to play with matches and the consequences if they do, how to be careful with smokers in the home, how to get a home heating system installed and make sure it's in good working condition, etc. These are some of the things children learn through community relations fire prevention programs. For this reason, focus should be on two main elements to get the job done: appropriate funds and adequate staff.

One of the goals I wanted to attain prior to retirement was to make fire prevention a bureau. Members of the Fire Prevention Bureau would not have to compete with firefighters for a promotion. Their tenure and ability to function in that division would be the criteria for promotion. In many cases, we think firefighters receive more benefits for working overtime because of their staffing shortage. There is a great difference between the salary of a member of the Fire Fighting Division and a member of any other division. When I was a Deputy Chief at the Training Division, I tried to bring some consistency in the pay scale. Since overtime could not be worked, I started a four-day workweek there. This gave members of the Fire Fighting Division incentive to be assigned to the Training Division. There were several obstacles encountered in this process. I was told that a form had to be developed, and that a different work schedule had to be created. These were nothing more than a variety of excuses to stop a positive effort to encourage members to work in divisions where overtime was minimal. If the plan to create incentives for members at the Training Academy was allowed to operate and be successful, it could have been instrumental in the detail of members to the Fire Fighting Division. Of course, it was unsuccessful just as many other ideas I attempted to accomplish. There were political angles and other obstacles from people who had no expertise in operating a Fire Department. Without a goal, you are out of control. If you don't know where you are going specifically, you may wind up anywhere. There are many people who will habitually criticize management. It does not make a difference what business you are in, it's positive thinking people who can make things happen.

Early in a firefighter's career, he or she begins to learn what management is all about. He/she can learn how to manage one's self, and from the commanding officer, you learn how to do it. With the help of the commanding officer and other members of the company, he/she will be able to complete this one-year probationary period. As he/she continues to work, they learn what is expected of them and start to emulate the actions of their Commanding Officers, as well as other successful firefighters in the unit. They can learn that for a promotion to a higher position, self-management must be developed. They must study the rules, regulations, and material that governs the operation of the Fire Department. They must read and study special orders, general orders, and any other official directives that are issued periodically. In essence, the method is the same for managing one's self as it is for learning how to use firefighting tools and equipment;

i.e., axes, hammers, ladders, etc. As a firefighter observes the different jobs performed by his Commanding Officer, he/she learns that they are in business of protecting the lives of the citizens, who they are committed to serve, as well as visitors to our great city. Six years have passed and now he/she is expecting to be an officer in the D.C. Fire Department. What they have learned as a firefighter can now be shared with others for greater productivity in the D.C. Fire Department.

In 1985, I hired a management consultant Group. This consultant was the same that had been hired by the District Government. My theory was perhaps that they could shed some light on issues that plagued the department, and could probably offer valuable recommendation to a solution. Originally, I was reluctant in employing this group, because I felt it would only have little positive impact, because of the difference among the members, Chief Officers, and labor. Many Chief Officers had not severed their ties between labor and management. This was one of the chief reasons that problems existed in obtaining support from the Captains and Battalion Fire Chiefs to adhere to departmental directives.

A strong affirmative action posture and the concept of merit promotion represented substantial changes in the D.C. Fire Department. Unfortunately, many Chief Officers viewed these changes as those which reflect the belief and style of the incumbent Fire Chief, and not as changes instituted by him in the role as Fire Chief to redirect the operation of the department. They neither accepted nor supported changes. They felt that by discrediting me as Fire Chief, either retirement or removal would be warranted. This would not allow any changes to take place. Some Chief Officers openly ridiculed the policies and actions of the Fire Chief in the presence of rank and file, or did other devious acts. For example, a Captain at Engine 8 had a sign on his truck that said, "The Fire Department is not only a job, it's a joke too." If the Captain of any unit is displaying that kind of mindset in behavior, the rank and file will do the same. I mentioned this action previously, in Chapter 9.

I have read some case law in reference to that very same type of situation in large organizations. The way they handled it was simple; the CEO was fired. The Fire Department can implement that same type of disciplinary action, but we have never gone that far because we never learned how to manage with the head and not the heart. Many officers would not implement directions that they felt unwarranted. They were to drill and enforce other regulations. Most of them interpreted any controversial decision by the Fire Chief in the worst possible light. Interpretations were made out of ignorance in terms of how the government works. Ignorance should be emphasized because if you don't know, you just don't know. And they blamed the Fire Chief for conditions not of his doing; e.g., the requested budget not being approved, the quality of recruits hired, and the kind of budgetary constraints put upon the department. To make matters worse, there was a lack of support from the Mayor's Office.

They were very upset because of senior officer promotions. As Fire Chief, it was made unequivocally clear from the beginning that in order to get promoted to Battalion Chief, you had to demonstrate the ability to operate as a company. Why should I give a member six companies to manage when he/she could not effectively manage one? Why should I give a member more responsibility when one has not demonstrated that one deserves the additional responsibility?

Many employees that I promoted were very instrumental in going into the community, sharing the goals and ideas of the Fire Chief. They also shared the Fire Chief's desire to work for the citizens of the District to eliminate and reduce fatalities caused by uncontrolled fires. There was some criticism made of the promotion of those senior officers. They were told on several occasions that most of the orders issued by the Fire Chief were either revised or rescinded, yet the analysis showed that it was only true for about five percent of the cases. They also told the group that the Fire Chief frequently reversed recommendations of the Trial Board, and Disciplinary Investigation Board. Of more than 110 cases before these boards, only one modification of a board's recommended action was found. That one was overturned because some members of the board tried to railroad the individual being tried. These are the type of actions by Chief Officers that would serve to discredit the Fire Chief, or make the proposed departmental changes appear questionable. These negative actions also contributed substantially to the overall tone of the department, which caused deterioration of employee morale. The group reported that unless these actions ceased, there would be a negative impact on the productivity of all areas of responsibility in the department, especially those areas that dealt with matters of life and death.

The group also talked about management issues and problems that had been identified. There was the issue of race. This issue resulted largely from discrimination of the department against black employees. There was a current perception by many that the department was now discriminating against whites, by showing favoritism towards blacks. Some changes that were implemented affected both races differently, and some members of both races just tended to be racist themselves. The racial issue was major, and needed to be resolved. The racism problem was one thing, and the management problem was another. Failure to utilize essential management tools in the operation of the Fire Department had nothing to do with race. However, what race did have something to do with was the subtle way it was used against blacks. Also, when talking with white firefighters, you would more than likely get the same answer. Basically, while Lieutenants, Captains, and Battalion Chiefs were members of the bargaining unit, talking about management had very little value in reference to success. As I have studied and learned, success is an inner concept. It is what you feel inside. Until Lieutenants, Captains, and Battalion Chiefs, start to think management inside, the Fire Department will never be totally successful in its endeavors.

Chapter 15

The Fire Department Budget

Immediately after I was appointed as permanent Assistant Fire Chief of the District of Columbia Fire Department in 1982, I knew that it was going to be a difficult task. I discovered early on how the budget process worked, so we prepared our budget based on our needs. Afterwards, it had to go to the Mayor's Budget Office, and we had to explain why we needed the amount of dollars being requested. We knew one thing for sure: The line item by line item was justified as close to the mark as possible, because there was no fat in the budget. You could not pad the budget because it was too easy to identify the need of the department. Nearly 95 percent of it was for personal services. We knew how many firefighters were on board; we knew how much they were being paid, including overtime; and we knew the number of unfunded positions, as well as vacancies. So, there was no excess that we could skim off the mark to make it work any other way; and I was not happy with some suggestions that were recommended.

After we met with the Mayor's budget staff, I went back to my office to meet with our overall staff, and discussed our budget mark, which was always 10-15 percent less than requested. I knew what the budget should be, and I was asked to tell the City Council that I could operate with less than I requested. That created a real problem for me, because in essence, I was being asked to misrepresent the truth, and it bothered me. I knew the chairman of the council knew the budget process, and had kept records on what we said by reviewing what we said last year as opposed to this year. They would want to know how we could operate with less than last year, then have to ask for a 20 million dollar supplement. This question did in fact come up, and I told them it would be extremely difficult.

We got into different segments of operations, but they would not give me the tools that I needed. For example, I requested 6 million dollars to re-build the Communication Division, and they gave me $300,000 and asked me to patch it up to see if we could last another year. In the mean time, when the Communications Division was unable to effectively dispatch, someone would call the media. Of course, I was the one who took the heat.

Until the city decides to give the Fire Chief resources needed to have efficiency in all services, you can anticipate having multiple problems. We were alleged to have a state-of-the-art Computer Aided Dispatch System in the Communications Division. However, we were still not dispatching the way we should. The Nation's Capital had the oldest, most run down communications system in the Washington Metropolitan Area.

The Mayor's budget director served all agencies within the District Government. This was our problem: They would cut our budget, and give us what they thought we should have. Each year the department would have to request additional funds to cover cost for salaries. Meetings with the Mayor's Budget Director were long and painful, and the director always won. The hearings frustrated me because everyone knew that it was practically impossible to operate a department with such budget con-straints. I was trapped in trying to operate with less than the adequate financial resources. Many times I felt myself to be in a no win situation, but I was determined to do my very best in running the department until I retired.

In 1983, I was sworn in as the Fire Chief, for the District of Columbia. I continuously tried working within the constraints of our budget of pre-vious years, after many months of having to navigate through the vari-ous divisions of the Fire Department, with the insufficient funds that were allocated. One department in particular, the Ambulance Service, had to do with the 10-15 percent less than requested budget wise. I was direct-ed to appear before a congressional hearing. I recall the Senator of New York, telling me that he saw that I was having a lot of problems with the Ambulance Service, as did the Chief before me. He asked me if I could tell him what was needed to have an efficient ambulance service. My reply was without hesitation, "We need more ambulances." In some way of ex-pression, he appeared surprised in my being so candid in my answer. He thanked me for my candor. I could not fully understand that, because I just stated an accuracy.

When I returned to my office, the direct lines from the City Administrator's Office were flashing. I answered and he told me that I could not say those kinds of things in a budget hearing at the Senate Office. At that point in time, I was baffled, because I did not know what he meant. I told him that I gave my professional statement of accuracy, in response to the senator's question. His response to me was that I could not tell the truth all the time, and that I had to think about my response, and not be so blunt. When we hung up our telephones, I reclined in my chair and thought, "What in the world is going on?" What other way could I have addressed the solution to the problem? It was obvious that they wanted me to misrepresent the facts, and support a budget that I didn't believe in.

In future budget sessions, I was very vocal in asserting the fact that we needed additional ambulances. I was extremely adamant in my demand for more ambulances or to be told why I could not get them. When we got into a situation where one needed to say what was appropriate according to others, we had to get the Mayor, the City Council, and everybody associated with the process because, in my case, I was not taking the fall.

Until the city decided to give the Fire Chief resources that would enable him to have efficiency in the Ambulance Service, you could anticipate having multiple problems. Due to the dispatching problem, the ambulances were not responding in the most timely and efficient manner. We were using a band aid approach to emergencies in sending $800,000 firetrucks to medical locals, when no ambulances were available. What if an injured person required transportation to a medical facility, were we going to transport with an $800,000 dollar firetruck? If the elected and appointed officials fail to provide an efficient ambulance service, the citizens should vigorously demand some answers.

I began thinking to myself, "Why am I being criticized so much about issues that I am not totally responsible for?" Those things run the whole gambit of problems that will never be resolved satisfactorily. Within my personal thinking, I ruled out paranoia.

In 1987, the campaign seemed to pick up to run T.R. Coleman out of the Fire Department. I had two Assistant Fire Chiefs, both white and male.

It had been the practice for the permanent Fire Chief to select an Assistant Chief.

The Assistant Chief selected was an outstanding officer. He was always on point in his duty performance. In our budget meetings, he was an amazing participant. He was not one that I would believe to make a questionable decision in the line of duty. For example, I had reason to go to the department motor pool. One of the mechanics said to me, "Chief, your ambulance is about ready. All we have to do now, is put a new set of tires on, and it will be ready to go." I asked him, who ordered the vehicle and for what. He said, "An Assistant Chief ordered it." I called the Assistant Chief into my office and asked him what the ambulance preparation was all about. He said that a Deputy Chief had made the request of him, in order to go to Richmond, Va., to pick up a military soldier who was a double amputee (paraplegic). I asked him if he was going to do that without the proper authorization. His response was, "I thought my position as Assistant Fire Chief, allowed me to make such a decision." I was so disappointed with his response, I sat silent for a period of time, because he had been an extraordinary kind of person. I began talking to him in the lowest tone of voice possible and still be audible. I asked him if he considered my position of authority as Fire Chief, in not bringing this to my attention. For at least an hour, I stressed the difficulty that the city would have faced if something of a serious nature would have happened. I said your and my careers with the Fire Department would have been in jeopardy. He spoke saying the ambulance request initially came from a member of the City Council. That being the case, I told him the Councilman should have discussed this with the City Mayor. I told the Assistant Chief that the ambulance was not to go beyond the confines of the city limits without my approval. Our demeanor in discussing this potential breach of department protocol in all aspects, was very civil. I reiterated to him some of the effects that could have derived from a catastrophic event. After he admitted that he made a mistake in judgment, I told him I agreed and that I would apprise the City Administrator of what was about to transpire. I then directed the Assistant Chief to return to his office, and to continue performing his assigned duties.

Approximately 2 months later, the Assistant Chief came to my office and advised me that he wanted to retire, and presented me his request. I asked him to sit down, and we had a lengthy discussion whereupon, I felt his desire was sincere. I told him that I would forward his papers through the proper channels for signature of approval immediately.

During the next week or so, an Assistant Chief came to my office. He said to me, "Chief, can I have a word with you for a few minutes if you're not busy." I said, "Sure have a seat. What's on your mind?" He said, "Chief, I don't think I am going to be able to roll with you much longer." I asked him what he meant. He said that he had been hearing about too much that was going on, and that he was too old to engage in some of the negativism that was being bounced around. He went on to say that he was considering retiring. I told him that he had some competent Deputy Chiefs that he could rely on, and that he should be able to perform his duties with ease. He said that was all well and good, but he didn't want to die on the job. He said, after all, that he had been in the department for over 30 years. I said, yep, that's true for both of us. We continued talking about some department issues and how they could best be resolved. Soon after, he went back to his office.

On Friday, a week later, the Assistant Chief came to my office, and presented me his retirement papers. I asked him if he was sure that was what he wanted to do. He said yes. I said ok and told him I would forward his papers through the proper channels. During that weekend, he called me and said that he had decided against retiring, and that he didn't know what in the hell he was thinking about. I called the Administrator's Office and left a message for him to call me. When he returned my call that day, I conveyed the Chief's request. He informed me that it was too late, that his retirement papers had been approved. I immediately relayed the news to the Assistant Chief.

I am confident that as you read this budget chapter, and how it tapered off into other subject matter, you will begin to connect the dots. I believe it may very well be that influences that were more sinister were in the equation. I might add however, anyone having the time and desire to retire if he or she chooses, should do so.

Chapter 16

Success of Operating Programs

The Fire Department ensures the safety of District of Columbia residents and visitors by preventing and extinguishing fires, as well as providing emergency ambulance service. The operating programs in the Fire Department, are supported by communications, supply, maintenance, training and administrative units capable of fully supporting the department's emergency operations. Radio communications between companies on the fireground will significantly improved the utilization of staff.

A new Communications System was developed for the Fire Department and the Emergency Ambulance Bureau. The ultimate goal of these efforts was to develop a service that was capable to meet the real emergency ambulance needs in the District. We had a consultant to study and determine what improvements were needed in the Fire Department's Communication system. The results gave us the most modern fire and ambulance communications system available.

It had already been determined that additional staff would be required for the system to increase the number of call takers and supervisors, as well as introduce quality control and improve maintenance. Equipment funds were requested to ensure that the present communications system was maintained until replaced.

There were at least four Spanish speaking dispatchers to accommodate the Latino residents of the District. There were 11 workstations authorized at fire alarm headquarters. Additional workstations were added, also, a third ambulance call taker and a second ambulance radio operator for the second frequency.

The procurement of a new communications system required the expertise of a specialist to oversee the analysis, design, procurement, and implementation of this system. In addition, two new positions were required for a computer specialist and technician to maintain the Computer Aided Dispatch (CAD) System. The computer specialist worked directly with the CAD system to maintain hardware and software, perform systems analysis, and enhance programs to meet specific needs of the Fire Department Command Center. The computer technicians were needed to assist with the CAD system maintenance and 24-hour system support. These positions were essential to the communications system upgrade that was underway.

The department increased the number of certain pieces of electronic equipment by 35 to 50 percent, because of the needs of the emergency ambulance service. Each ambulance carried an additional radio. More portable radios were in use. Cardiac telemetry equipment was added for advanced life support. Electronic sirens were placed on many pieces of apparatus.

The communications system consisted of a computer aided dispatch system, various radio systems, enhanced 911 equipment, status maps, a fire box alarm system, a vocal alarm system, a telephone system recording. The Fire Department requested purchase of a fire and ambulance radio system with computer aided dispatch and ancillary equipment. The installation of this system required renovation at the communication center, and at satellite antenna locations. This was the first part of a three-year project to upgrade the department's fire and ambulance communications system.

In 1987, it was recognized that the Fire Department's radio and communications systems were obsolete and the equipment was over aged. Not enough frequencies were available for communications between units and command on the scene at fires and other emergencies to handle the growing number of ambulance calls. Up to 73 percent of some types of radios were over aged. In addition, the computer aided dispatch (CAD) system was 10 years old and technically inadequate because programming had reached the capacity of its memory, making programs enhancements impossible.

In the fiscal year 1988 operating budget, the District authorized for a consultant to design and supervise the installation of a new fire and ambulance communications system with a new computer aided dispatch. In order to provide additional radio frequencies, all radios were converted to frequencies in the 800-mega hertz range.

Two new features were included in the proposed modernization digital status keeping, and automatic vehicle location. Digital status keeping coded standard messages between units, and headquarters so a unit could transmit them with the touch of a button and have them automatically recorded at headquarters. Two advantages of this was to cut down the air time to transmit messages and to more accurately record response times for units. Automatic vehicle location for the emergency ambulance service allowed dispatchers to pick the ambulance that was actually closest to an emergency rather than relying on preplanned dispositions to choose which unit to send.

The CAD system was replaced with a new computer, programmed and supported to provide more efficient dispatching and detailed data for analysis of response experience. Programming and training for this system was part of the project. This project greatly improved the District's control management and utilization of its fire protection and emergency ambulance resources. The new computer aided dispatch system provided a database for more efficient unit location to allow more expeditious response to emergencies, and for detailed analysis of response experience. Emergency incident managers had adequate radio frequencies so they could make maximum use of the resources available to contain the emergency. The system was necessary to ensure that the community received the greatest degrees of protection.

Chapter 17

The Ambulance Crisis

The so-called "Ambulance Crisis" caused many unpleasant moments for me and members of my immediate staff. The ambulance problems didn't just happen overnight, during my tenure as Fire Chief. They began many years ago prior to my appointment to the Fire Department in 1953. Many of my critics didn't know that the District of Columbia had an ambulance service in 1953. However, there were specific authorizations for the functioning of Fire Department and Ambulance Service in the District of Columbia. The City Commissioners at the time established the Emergency Ambulance Service on September 6, 1957, by Order No. 57-1667. It read thus, "There is hereby established in the District of Columbia an Emergency Ambulance Service to consist of emergency ambulance vehicles, equipment, and crews of the D. C. Fire Department and Emergency Ambulance Units. The Fire Department shall have coordinating supervision of the Emergency Ambulance Service, including the operation of a central dispatching service for emergency ambu-lances and equipment, and crews assigned to the service. The Fire Department may enter into agreements with other District Departments to obtain reimbursements for services, such departments in connection with the Fire Department's coordination, operation, and supervision of the Emergency Ambulance Service."

On the same day, an amendment, Reorganization Order #38, redefined the purpose of the Fire Department: "The Fire Department is established for the purpose of providing the maxima protection of life and property in the community with the particular reference to the prevention of fires before they occur, and to expeditiously extinguishing of fires after they occur, and also with reference to providing various emergency services in connection with protecting life and property. The Fire Department shall have the coordinating supervision over the District of Columbia's Ambulance Service."

The District of Columbia hired the first career medically experienced civilian members to the Emergency Ambulance Service on October 15, 1974. On July 3, 1976, the first paramedic class of the District of Columbia, conducted at Georgetown University, graduated 16 members of the Emergency Ambulance Service. However, it wasn't until September 1977, that the first Mobile Intensive Core Unit (MICU), Mobile 25, was placed into service.

The Mayor of the District of Columbia issued Executive Order No. 81-176 on November 9, 1981, establishing the Emergency Ambulance Service as a Division within the District of Columbia Fire Department: "The Fire Department is established for the purpose of providing the maximum protection of life and property in the community, and coordinating supervision over the Emergency Ambulance Division which shall provide basic and advanced life support 24 hours a day in cases of acute illness and injury to the residents and visitors of the District of Columbia."

One of my critics, who was a member of the City Council, stated that I was a good Fire Chief, but couldn't manage the ambulance service. You may think a man making a statement at that time should not be representing members of his ward, because, if he had any knowledge of management, he would know that management is management. Whether you are managing a mom and pop grocery store or the President of General Motors, the results would be the same. It appeared that some members of the City Council made a consistent practice of recommending that the Mayor ask for my resignation or remove the Ambulance Bureau from under my command. These individuals made a mockery of the budget process by asking unintelligent questions that their staff members probably wrote for them. It was evident that the questions asked were predicated on what was written and on the electronic media. It was appalling that those same members, who sat there like "bumps on a stump," criticized me for not fixing the Ambulance Service, when they knew specifically what it would take to operate any ambulance service. The answer was resources, money to purchase more units, equipment, and additional trained personnel. It was that simple. You didn't have to be highly intelligent to understand that.

In my opinion, many members of the City Council were not really concerned about the level of service rendered to the citizens of this city. They were just trying to exacerbate the problem in a public forum at my expense. I think they were smart enough to realize that each year the price tag for any service rendered in the District of Columbia continued to climb. Did they recommend adding funds to my already strapped budget? "No." They just hopped on the bandwagon smiling like "the cat that just swallowed the canary," that I had been defeated. The City Council has a responsibility to ensure that the funds are made available for services to the citizens of the District of Columbia. I am very appreciative for the support that I received from Attorney Wilhelmina Rolark, the Chairperson of the Judiciary Committee having budget oversight for the Fire Department. Funds were made available to purchase two additional ambulances. There was no reason to buy any more than two, because at the time, we did not have staff required to place them in service. For each of the 6 years that I was Fire Chief, I requested funds to increase our fleet. We had information available, which stated that with the number of responses made each year, 35 ambulances would be required. We also requested funding to replace the 13-year old Computer Aided Dispatch System (CADS), at communications for a price at that time of 6 million dollars. Today the price has escalated to 25 million dollars, and the system would be obsolete in 4 or 5 years. For Computer Aided Dispatch System, manufacturers can't keep up with the changes in new technology. When I made those requests, the response was always the same, "No money in the budget to fund it." In relation to communications, an approval of $400,000 was to try to keep it working another year, and maybe then funds would be available. In my first Congressional Budget hearing, the Chairman of the Budget Subcommittee, talked to me for about an hour at the hearing. He asked me, "Chief, what do you think is needed to fill vacancies and increase the number of ambulances available for emergencies?" My answer was and would be the same today as it was then, hire people in the same way we hire firefighters, that is to give a test to the applicants having at least a high school education, train them at the required standards, and assign them to the service. That was very simple. He thanked me for my candor. It was stated by members of the Mayor's Medical Advisory Committee and doctors at various hospitals, that if the department adopted that type of hiring practice, the service would be inadequate. Therefore, the recommendation was rejected. As a result, we continue to have problems using the system that was in place at that time. You just couldn't get enough people qualified under the requirements that existed. The requirement was 4 years of combined technical experience in some kind of medical field. There had been many recommendations submitted related to operating and managing an ambulance service.

By 1960, the Ambulance Service, which it was called at that time, had seven ambulances to serve a population of over 600,000 people. I was a member of Engine 10 before being assigned to the First Battalion. The First Battalion had Ambulance 5, NE, and 7, SE, which responded to more emergencies than all the other units combined. I performed duties on both of those ambulances for more than 4 years. I knew from experience what was needed to fix it to the point of where we would be efficient and proud, because of the experience I received, not only from a management point-of-view, but from the ground up, I knew how to handle every segment of the Ambulance Division.

According to the newspaper, an article called on the Mayor to remove the ambulance service from the jurisdiction of the Fire Department, citing a year of failure to correct management deficiencies which continued to produce unacceptable life threatening lapses in emergency responses, and that he put the ambulance service under the leadership of a professional director in Public Health Services. If the chairman of the council had conducted a research, the council members may have remembered that Public Health Services, is where it came from, because of an inefficient operation. The Public Health Commissioner was strapped with all the problems that existed. All of the supporters of this removal recommendation should have been asked what criteria were used in making the unanimous decision. The Mayor was pressured from every possible angle to make some adjustments to move the ambulance service from under my direct control. I understand clearly why the Mayor made the change of having the ambulance personnel to report directly to the newly appointed City Administrator. He was tired of having me take the weight for something over which I had no control, to manage the ambulance service with inadequate funding. All of the members of the Judiciary Committee were aware of this. The Mayor's wisdom knew the reason for all this scrutiny, in reference to the Ambulance Service. He knew when I started promoting Blacks to management positions, he had to fight all kinds of opposition. The Mayor also knew that some undermining elements were to get rid of Coleman, nothing else.

If anyone had a problem with reporting to the Fire Chief, that person should have been fired, or told at the very beginning that the Fire Chief was totally in charge. But that didn't happen. A year and a half past, the reporting system took place and with all the money and time spent, the Fire Chief called the City Administrator, Personnel Director, Police Department, and many other agency heads. I spent many hours trying to submit recommendations, and agree on some operational systems that would enhance the operation. Up to this point, nothing had been done in addition to what I had already done prior to my retirement. One thing that was always appalling to me was why was I being criticized, when the Assistant Fire Chief of Operations had direct day-to-day operational control over the Ambulance Service. My Operations Chief was white. He wasn't the one they wanted to get rid of. Why shouldn't they attack him?

If the media had some problem with the Ambulance Service, they should have gone and talked with the Operations Chief, because the Operations Chief reported to me on situations in reference to his Bureau. So why all this information floating around about me fouling up the Ambulance Service. It appears to me that the Operations Chief should have been mentioned in there somewhere, because he had direct responsibility for the day-to-day operation of the Ambulance Service. In the past, the Fire Chief, and other members of the District Government, had taken trips to various cities to try and culminate information in hopes of returning home with an operations manual for the Ambulance Service for the District of Columbia. I have talked about this consistently with others, asking why the people who get involved don't understand that the District of Columbia is a unique city within itself. The only thing that I believe they look at when they go to other cities is the population and the geographical territory, being about the same. What about the people? Do the people use the Ambulance Service the way they do in the District of Columbia? Do the people in these other cities have to pay for service? All these elements are contributing factors to the success of an Ambulance Service. The Ambulance Service in the District of Columbia was free at one time. People are always reluctant to change. A major improvement could have been made in the Emergency Service by revamping the dispatching policies. Unlike every other area jurisdiction, the District seldom sent the nearest available unit to a major emergency scene. There were times when a victim had to wait a time for an Advance Life Support Unit to come from a great distance.

Chronology: D.C. Ambulance Service

1957 - Independently operated by D.C. Health and Fire Departments, and local hospitals. Ambulances were manned by doctors, interns, and physician's assistants (7 Ambulances and 23,000 Incidents).

1957 - Consolidated under the control of D.C. Fire Department. The Emergency Ambulance Service became a section of the Firefighting Division (7 Ambulances and 24,000 Incidents).

1968 - Call volumes doubled in the 11 years since D.C. Fire Department assumed control of the Ambulance Service, and only one additional unit was placed in service (8 Ambulances and 48,000 Incidents).

1973 - Call volumes rose by another 50 percent. Firefighters began publically voicing their displeasure about ambulance duty in general, long waiting times in hospital emergency rooms, and the high ratio of non-emergency calls (10 Ambulances and 72,000 Incidents).

1974 - Public controversy was resolved with the decision to leave the Ambulance Service in the Fire Department, in a civilian capacity. Career EMT's were hired and Firefighters were returned to fire duty (10 Ambulances and 75,000 Incidents).

1976 - First Paramedic class in Washington, D.C., was graduated from Georgetown University. First "Advance Life Support Unit" placed in service (12 Ambulances and 76,000 Incidents).

1981 - Ambulance Service recognized and elevated from a Section in the Firefighting Division to the Emergency Ambulance Division (14 Ambulances and 93,000 Incidents).

1987 - Ambulance Service upgrade from a Division to the Emergency Ambulance Bureau. Ambulance Director's position, by Executive Order on May 6 by the Mayor, was made equal to the rank of an Assistant Fire Chief. Danny R. Mott was made Director equal to Deputy Chief (21 Ambulances and 115,000 Incidents).

1988 - The Mayor appoints a Steering Committee to research and recommend an EMS System Design (21 Ambulances and 125,000 Incidents).

In life threatening situations when no ambulances were available, the rescue squad would transport the patient to the closest appropriate hospital. However; the rescue squad vehicle is the last resort for any transport.

We should have done away with priority dispatch, because in my opinion, it had never worked in the Fire Department. I fought this quite vigorously. There were people closer to the Mayor who saw to it that the Mayor agreed to put the Priority Dispatch System into place. Implementing that system created a lot of problems for members of the Communication Division. Many of the problems that existed related to late responses. The run getting lost was related to the system of priority dispatch. The call taker would call, and it would be given to a dispatcher, and the dispatcher would determine the level of priority, predicated on the information received. The caller needed to get a response right away or he would have to wait for a half hour. A few times they actually got lost, never did dispatch. I think the best operational process would have been to get the call and to dispatch it right away; that way you would know that it is gone. But, if one got the call, and gave it to another person, that person gave it to somebody else, and someone else determined what the priority was, it would have been easy for a system like that to get fouled up. It was just unreal having all these people involved in the Ambulance Service. In other area jurisdictions, it was policy that the nearest available unit, either an engine company, a truck company, or a rescue squad, would have been sent to render immediate basic aid, while the Mobile Intensive Care Unit was enroute. However, in the District, such a response was not automatic. It was a laxed enforced basic order for the Communications Division. They simply didn't know what they were doing under the circumstances. Those people were doing the best job they could with the equipment available. It just seems that the Nation's Capital would have the state-of-the-art equipment from a Computer Aided Dispatch System to the Vehicle Locators. All the information is out here. You can buy it, but if you don't want to buy it, then don't go around blaming somebody else for not getting the job done. Getting the job done was really not the issue. Reporters were looking for a story to justify their mere existence. The D.C. Ambulance Service was always over worked, and even under staffed, until the day I retired. At one time, it had only two Mobile Intensive Care Units to cover the entire city. The system was always seriously abused by many people who used it like a taxi service for clinic appointments and other non-emergencies. Consequently, the reality was that the Fire Department's Communication Division used all of the life support units available; therefore, making it mandatory to send the nearest available emergency unit to assist on major life threatening calls. There was a statement made in writing by a newspaper columnist that the numbers tell the tale that EMS had become the overwhelming majority of call runs by most Fire Departments in the country, out numbering fire runs by 5 to 1, in some communities. While EMS was increasingly perceived by the public as the priority role of the fireservice then, tradition and politics often prevented EMS from getting the kind of attention and priority it needed within the fireservice itself. A survey was conducted that indicated some dramatic figures to the extent that the EMS dominated the service of the Fire Departments. For example, in Chicago, Ill., there were 59,172 fire calls, but 161,231 EMS runs. In Los Angeles, Calif., there were 92,155 fire calls, and 158,613 EMS responses. In Washington, D.C., there were 19,666 fire alarms, and 92,855 EMS calls. The EMS runs out-numbered fire calls by more than 4 to 1. The EMS Division accounted for over 9 percent of the total Fire Department budget.

There has for sometime been a disparity in the work done by EMS, and the support and acceptance within the hierarchy of the fireservices, the fact that it was relatively new within the fireservice. For many years, there was no field Health Care, except the basic first aid, which was limited to assisting injured firefighters at the scene. Service to all others was by hospital-based ambulances, private services, funeral home ambulances, and in some cases, no pre-hospital care at all.

In early 1970's federal funds were provided to the EMS Service, through the Highway Safety Act, and the Fire Department reluctantly began providing some limited basic life support, mainly to ensure that communities would get some federal funding. This brought about an aggressive effort to get the EMS going. The Fire Department was forced into accepting the role of Emergency Medical Providers, a role that they did not like. Firefighters were ordered to take EMT classes, assigned to ambulance duty, causing their traditional routines to be disrupted. All of this brought about negative feelings toward EMS, many of which remained until my retirement.

The political role played in advancing the EMS into traditional realm of the firefighter, by the lure of federal money, pushed many local governments into expanding their services into EMS, well before they, and their communities were ready to do the job. The political tide was difficult to hold back once it got started. Free ambulance service given to a particular community, immediately, demands came from others wanting the same. Then came the Advanced Life Support glamorized by the news media in emergency rescue situations. The political pressures from upper levels filtered down to the Fire Chiefs, who came under extreme pressure in taking an EMS role with little or no knowledge of how to proceed in the process. It has brought about priority spending of financial resources between EMS, and fire suppression. This put the EMS in a bad light with the public, which in turn had a tremendous negative impact on me as Fire Chief. The facts are that we were poorly staffed, ill equipped, and slow to respond. This in itself indicated that there needed to be more identifiable mission departments, since managers in the local department of Government would be asking the Fire Department to do new, non-traditional things.

There was an incident that occurred, and it was very troubling to me. It had to do with a man who supposedly had a short time to live as a result of AIDS. He called me at my office, and stated that I should step down as the Fire Chief. I asked him why. He responded saying that I was messing up the Ambulance Service, and if his parents came to this city and needed an ambulance, he had no confidence that they would get one.

I explained to him that improvements were being made in the Ambulance Service, and that he could rest assured that his family would get adequate emergency service if needed. He replied by saying he couldn't buy it; also, he stated that I had ten days to step down and retire, or he was going to make his death occur at my home's front yard.

A prayer vigil was established by Dr. Calvin W. Rolark, Sr., a person who knew what the system was doing to me. When this occurred, I had been called to a meeting in the Mayor's Office that morning. I was informed that a number of clergy leaders were present, along with well over one hundred citizens, in support of this effort. I understand that some calls had been made to Dr. Rolark to call the vigil off. Also, I understand that a Council Member showed up at the vigil, accompanied by the man who threatened to end his life in front of my house.

A concerted effort was made in adjusting the close monitoring of Fire and EMS Communications in upgrading the dispatching facilities. It was extremely important to ensure that the dispatching personnel assigned to EMS were EMS qualified. It was realized early on that many of the best EMS personnel had limited interest in firefighting, and many of the best firefighters had limited interest in EMS. Therefore, ways had to be structured to allow for advancement beyond, say, a paramedic or possibly shift supervisor. The fact that EMS is one of the most stressful possible professions, field personnel ought to have been allowed the opportunity to strive for supervisory and top management positions, as well as clinical assignments. Anything less, made employees less than enthusiastic about what they are assigned to do, knowing that they may be stuck in a dead-end job. A career ladder needed to be structured that would allow personnel to be trained in management, business, and computer training courses. The EMS staff providers are required by state law to keep their skills proficient and pass regular state test to prove their competence. With all public expectations of EMS personnel, the responsibility of the Fire Departments needed to have a reality check, because the EMS was their primary public service.

I have cited one segment of the many problems that plagued the Ambulance Service in the District of Columbia. The news media have the propensity for reporting any and all negative sides of any given situation. One of my Acting Deputy Fire Chiefs forwarded me a memorandum concerning media coverage. He wrote, "Somewhere in the system, we lack positive media coverage, for I am constantly being reminded of this division's mistakes in the newspaper. The reason I bring this to your attention, I was personally involved in the following incidents: Media Nine, Sousa Bridge, Auto Accident, Driver D.O.A.; Fire 915 M St., NW., transported one person, 1700 Block of Columbia Road, NW., Fireman transported, Fire 2000-415 St. NE., three persons transported, two critical. Not to mention the numerous amount of responses and overdoses. I would not write this report if I didn't believe that we must receive positive media coverage and report what we are doing for the citizens we serve." The news media cannot be held totally responsible for printing information concerning the alleged deficiencies of the Ambulance Service. Information had to have been given to them by someone inside the organization. Some of the members of the Fire Department had personal vendettas against me and my administration. They were prone to discredit anything that I did.

Some of the people assigned to Communication knew exactly how to get information out. What would happen is that when the Ambulance Service received a call, these people would call their contact at the local newspapers and give them all the information available. That's why so many times when the ambulance crew arrived on the scene, the television cameras were already there. This couldn't have happened unless someone on the inside was doing it. With disgruntled employees, there must be a system in place to relieve these people from their duties. The Mayor or the City Manager needed to sit down with the Fire Chief and his staff to prepare an operational procedure for the Fire Department in this area to take into consideration the whole operation as it relates to members of that division, and when this procedure was done, make sure that everybody in the organization understood clearly that it was an emergency operation. When they were not in compliance with the established procedures, they were actually jeopardizing somebody's life that they were sworn to protect. Members of the D.C. Fire Department had gotten so complacent with breaking rules and regulations until it was pathetic, because they believed that the union would take care of them. The union could not take care of you when you were in violation of a rule, regulation, or order. The only thing the establishment had to do was create a disciplinary manual that simply said, "When you are not in compliance, you will be charged accordingly." When you are not afraid of having to pay for a crime, one has to understand that everybody has to pay; that's the way the system should be. You just can't go through life doing what you want to do, when you want to do it, and not have to pay. If an individual does not come to work on time, he should have to pay. There should have been policies in the Fire Department that simply said, "There is no excuse for tardiness." If you know you have to be at work, then you make the necessary arrangements to be there. When you get into a situation where a firehouse has a different set of rules than you have in another house, you have a problem. If the Captain didn't care about his people coming to work late, he hadn't made a management decision that simply said, "When you are late, you don't work."

When the Captains, Lieutenants, and Sergeants are all members of the union, that should not be. You cannot be in labor and management at the same time, because if you are paying dues to the labor organization, you are part of it. I had Battalion Chiefs, Deputy Chiefs, Assistant Chiefs, and people who were in charge of 1,300 to 1,400 people, and members of the labor union, to say that they just didn't get out because of the benefits they received when they were in the union. I was told by a member, who was responsible for negotiating the contract for the City, that I, as Fire Chief, could not ask him if he belonged to the union. These were some of the things that created real problems for me. I was steadfast in my belief. I believed that the Fire Chief and management were in charge, and if you didn't show respect in relation to getting your job done, I would take appropriate action. There were too many levels of non-management. The Lieutenant would feel that he was management; but who was he supporting? He was supporting labor. The Captains supported labor. The Battalion Chiefs supported labor. Well, who was supporting management?

Here is an incident that could be one of many that a Fire Chief may be confronted with, when personnel choose not to adhere to established rules, and regulations:

In one of our city firehouses, there was a Captain, who took it upon himself to paint the firehouse he was in charge of, the color of the football team, burgundy, as well as displaying their logo, on the front of the firehouse. An Assistant Fire Chief was making the routine inspection of firehouses and observed the physical activity going on in front of the building, and directed the Captain to have the logo removed. The Assistant Chief informed me as to what he observed, and the action he took. As Fire Chief, I concurred, and directed him to file an official report.

It appears that the Captain got the material supply department involved. This was obviously outside the direct authority of the Captain, to do this. In other words, this was not cleared through any official in upper management. This incident began involving others, such as the union that represents their members in the Fire Department. They are believed to have contacted the City Mayor. He in turn notified me to meet him at Number 8 firehouse. We met, had a brief discussion, and disagreed on certain specifics. However, he directed me to have the logo restored to the front of the firehouse, and I complied.

Some members of my staff and firemen verbalized that the Mayor upstaged me by caving in to the union and Captain of the firehouse. However, in some way, I felt that the Mayor may have been pulled into this act of insubordination without examining the facts. It appears that he used his position to ameliorate a discourse involving this matter, and others, that often flared up involving staff that objected to my management style.

This particular incident could have developed into an incendiary type personal reaction, as it involved the football team. The color of paint used did not match any other firehouse in the city, which could have caused many questions to be asked by residents in the city. I believe the Mayor, to be a staunch supporter of the football team, and I am equally. However, the management style infrastructure must always be examined, and strengthened where necessary, to be certain of avoidance to the firehouse paint/ logo episode.

Chapter 18

Snow Episode

I believe that everyone in this country and everywhere else heard on the radio, saw on television, and read in all the newspapers how I ordered the firefighters to remove snow that had been piled up on firehouse lawns and thrown in the streets. What is significant about this situation is the newscasters were unaware of what precipitated the order and its execution. The city had been grossly criticized over the handling of a snowstorm that crippled the entire city for weeks.

The Mayor was criticized for being out of town during that period, as though he knew when it was going to snow. After the Mayor had returned to the city, he set his sights on getting the District Government to work in harmony and combat the snowfall. As the Mayor rode through the city, he noticed some firestations had been cleared of snow from the sidewalks and aprons, and piled up near the buildings. Clearing the snow and ice from the center of the street, roadway, and sidewalks adjacent to the firestations had always been a part of our job description. If we failed to accomplish that mission in its entirety, people would have had problems walking to the firestations to notify the Fire Department of any emergencies in our vicinity.

Following the snowfall, I received a call from my communications center to contact the City Administrator. I complied with the directive, and I was told by the City Administrator that the Mayor was riding around, and saw that snow had not been removed from the streets and sidewalks at Engine 7, which was located at Half and M Streets, S.W. A threat was relayed to me through the City Administrator, supposedly, that if the Mayor saw that snow had not been removed at a later time, "heads were going to roll." I had no knowledge as to what he meant by "heads were going to roll," however, I felt that this was totally disrespectful, and unwarranted, to say the least. However, I questioned the veracity of such as coming from the Mayor. Having listened to the City Administrator relaying the Mayor's message, I immediately called the on-duty Deputy Fire Chief, and ordered him to meet me at Engine 7 at 7th & Half St. SW., at the repair shop.

The two companies of firefighters had cleared the snow from the northwest corner of Half Street, S.W. to First Street, and North on Half Street, to the inspection station, a distance of approximately 300 yards. I was very unhappy about that situation. However, the conditions surrounding my position left me with no choice at that time.

The Sunday following that incident was the straw that broke the camel's back. About three or four o'clock that afternoon, the Mayor called and said he noticed that some companies had not removed the snow as ordered. I was somewhat appalled at the beginning of his conversation, because I thought the Deputy Fire Chief had not complied with my orders to ensure all sidewalks and roadways adjacent to the firestations had been cleared of snow and ice. As the Mayor continued, I questioned the intent and the reason for the additional concern. He said he wanted me to personally and physically ensure that the snow piled up at every firestation in the city be removed and thrown in the streets to be dissipated by the cars as they traveled in those locations.

I told him that this was a bad move, and if I gave that order, we would get into serious trouble with the rank and file firefighters. I also stated, "You must understand Mr. Mayor, as soon as I visit the first firestation, the news media will be all over our case." His reply was, "I don't care, they should have some pride in the way the area looks surrounding the firestations." I reiterated to the Mayor that that was not the way to address the situation. His reply was, "Do it." I was furious in not being able to persuade the Mayor to rescind his order. I was at home when I received the order, and pondered my options on how to have the Mayor rethink his order, and not be insubordinate. However, I set out to comply with the Mayor's directive. Prior to reaching my firststation, it started to rain. I thought I would try to persuade the Mayor once again to rescind the order. The Mayor's Command Center called his house; the Mayor did not answer. I wanted the mayor to know that it had started to rain, and the rain would compound the problem of restoring the snow back on the firestation lawn. I could see it causing ridicule as clearly as your nose on your face, via the news media, which in fact accosted the Mayor, at Engine 19, and filmed him shoveling snow from the firehouse sidewalk/lawn into the street.

The saga continued the remainder of the evening, and aired everyday for the next 2 weeks on local television stations. Some firefighters were alleged to have been injured while they performed the snow shoveling duties. As a result, an engine company was placed out of service. It was believed that the claim of injury was part of a conspiracy to cast reflection on management's decision-making; and they were successful. A photo showed a group of men wearing white tee shirts, with red lettering that read, "D.C. Fire Department Snow Shovel Drill Team." They also wore firemen style red helmets. The picture showed them marching to a form of cadence, pushing snow shovels along the road as if shoveling snow. The men followed behind a moving firetruck.

In my opinion, this was a discreditable act that none of our loyal, and otherwise dedicated firefighters deserved. They did not deserve to share the criticism for upper management and city officials' orders. I have lived with that order, but it affected my family to such degree that my wife Uvaghn asked me, "Why are you continuing to put up with this B.S. when you have enough time to retire?" Many a person, including my immediate family, didn't know that I had a responsibility of added dimension that I chose to address before retiring. Anytime an individual sets out to attempt to address perceived notions of sorts that impact on subordinates and their mission that is in need of change, he must continue even though he anticipates great adversities.

The Fire Department for many years had been the advocate of discrimination, to the point where blacks had little or no chance for advancement to high-level management positions. I felt that I wanted to dismantle the effects of the "old boy system" of promoting, and I had not completed that objective, so, I had to knuckle down and bow out to criticism in order to keep going until such time as I decided that I was ready to retire.

To add insult to injury, there were some problems with the Public Works Department. The trucks with shovels on the front had no regards for fire hydrants when pushing snow out of the streets. They just covered them up. So, they created another dilemma for the Fire Department, which meant that the firefighters had to go out in the street and remove the snow from around the fire hydrants in order to gain access when needed. I felt that the concern for my firefighters had not been demonstrated by the Department of Public Works. I addressed the issue with the Mayor regardless of what some employee representatives said or their reason for saying it. I cared about my employees the same as I was concerned about the citizens in giving them the service that they so rightfully deserved.

Because of the system, and the way that the news was filtered throughout the city, it would leave one to believe that the Fire Chief had no concern for his employees. It is almost impossible for a Fire Chief to maintain the kind of respect that is needed to perform his duties in a most efficient manner when unsubstantiated statements are made. Fire Chiefs need to feel that the citizens of the community are pleased with the service they render, with the reflection of his/her firefighters performing their duties.

So as the "snow saga" continued, would you believe that we had another snow fall on top of the one I had been so grossly criticized? That morning when I got up to go to work, I noticed an accumulation of snow had fallen during the night. I had not been listening to the radio to find out information in reference to school closings, or the effect it had on the Federal and District Government employees. I had my mind focused on my Fire Department radio. In doing so, I left the house prior to the announcement that the District Government was closed other than for people who were identified as "essential employees."

I went out and started my official car. I had put the chains on over the snow grips so that I could make it to my office, but it was too much snow, so I became stuck in my car. I contacted my Communications Division, and directed them to place Truck 11 on the air, and to remove the snow from around my car so that I could go to my office. I was probably the only person at my office most of the day, because most of my staff had heard the announcement. I was criticized for having the firefighters shovel my official car out of the snow so that I could go to work. I think that was one of the most ridiculous articles that the news media could have printed. As Fire Chief, who am I going to call to get me out of the snow? If there is a situation that arises with the Battalion Fire Chief, so to speak, that was stuck in the snow, who would my Battalion Fire Chief call to get him out of the snow? He would call "firefighters." The Fire Chief cannot do it, because that's another element of criticism that the news media wanted to use as related to "character assassination."

I asked some agency personnel what they would have done, and nearly all agreed that my action was appropriate. Up until today, you could ask anyone of my staff, "If you were the Fire Chief and got stuck in the snow, who would you call?" They would say, "Call the firefighters just as they are called to an accident scene," oftentimes no matter how minor the accident.

Chapter 19

No Confidence Vote of the Fire Chief

I received a vote of no confidence from the local union on my performance as the manager of the Fire Department. The local claims that the vote cast was 180 to 2, showing dissatisfaction with my management in the first 9 months of my administration. The president of the local union said the vote overrode an earlier decision by the Union's Executive Board to continue to work with me to overcome differences, and not resort to a vote of no confidence.

Minutes before the meeting that night, it was said that a small group of firefighters gathered in the parking lot of their Union Hall in North East, Washington. The topic was about me. One firefighter said, "I know which way I will vote, and further, I just don't think he's qualified. He's just a puppet for the Mayor, and I don't understand why." The members in the group agreed unanimously, entered the hall, and cast their vote.

My response to that is I felt it to be very unfortunate that the union continued to refuse to cooperate with me as Chief of the Department. It seems that the vote came as a result of my unwillingness to bend to the wishes of the union that had constantly opposed me in almost all of my efforts.

The vote of no confidence marked the first such action in the 48 years of the union's memory, according to its president at that time. At this time, it is not clear if anyone had tried to ascertain why the no confidence vote was cast, or who presented this initial motion for such action to be taken. In my opinion as to why, for many years the union has virtually controlled the Fire Chiefs, with a threat of such a vote; however as I stated earlier, to my understanding this was the first time it had happened. I understand that many firefighters, active and retired, were aggressively pursuing members of the news media for their version of my management style and ability as head of the District of Columbia Fire Department.

I understand that a retired firefighter made a statement to the media saying that he was in complete agreement with the local union no confidence vote. He stated that I was trying to do my job the best, in order to make the mayor look good, when my priorities should have been the protection of the citizens of the District. That part of the statement to the media is true. If my performance is at the level of proficiency established by the Mayor, everybody benefits in some way.

I had some knowledge of the intent to cast the vote of no confidence several days before it happened; however, I refused to agree with the union's demands. I have attended union meetings in the past wherein I realized the president had very little control over the beer drinking and obnoxious behaving members. At the time of the vote, it is unknown for sure how many members were present, white or black.

The Local Union has always opposed anyone who spoke of equality, let alone made it happen. I have never had an interest in discrimination against white people in my profession. My focus was to create a racial balance within the department, which had not been done before. It appeared that things were going fairly well, until I began promoting black firefighters out of rank order, as the names appeared on the promotional register. However, the procedure was within the guidelines established in an agreement with the union's contract. Black firefighters were promoted to middle and top level positions as vacancies occurred, based on qualifications, and the need of service.

The union's white firefighters objected to my development of a program to place more blacks in the department's upper leadership positions. Also, they felt that I was moving too fast in an effort to enhance black's promotional opportunities. I made a vow to continue to promote blacks until upper ranks were equalized. The union's actions underscored its opposition to affirmative action as it related to blacks.

I believe my actions made some people unhappy, when it reflects on racism. I wanted there to be no mistake about my stance, no matter how unpleasant it applied to anyone. In the past, we read and experienced the actions of some top-level management people in the Fire Department, who took a stand against injustice and inequality. What appeared to suggest preferential treatment given to black firemen on all rank levels was not without anticipation that white firemen would be resentful.

To put the accusation more perspicuous, there was a suit filed against the District of Columbia Government by the Progressive Firefighters Association for alleged discrimination in the overall hiring practices of its rank and file personnel. As the suit progressed in the legal system, it was discerned that the need to fill the Battalion Chief positions were of a dire necessity. It was agreed by all parties that the suit had no bearing on this particular position. There were 20 slots in the Battalion Chief position that needed to be filled. As a result, 13 white and 7 black officers were promoted. The union members were up in arms, and otherwise outraged, because they felt that more whites should have been promoted within those 20 slots. However, my selection decision prevailed as a result of a reality check.

During my tenure in the District of Columbia Fire Department, I was credited for challenging a number of administrative practices, similar to the concerns by Lieutenants, Captains, and Battalion Chiefs. The administrative practices were discouraging to black firefighters, which inhibited their chances of upward mobility. I continually observed the work performance of firemen who I believed demonstrated leadership qualities.

Listed below are some of the firemen that were promoted. A few have since retired and some have moved on to leadership positions in other states' Fire Departments. I learned this by telephone calls from them, and by attending some special occasions involving firemen.

- Everett A. (Benny) Green was appointed to the D.C. Fire Department on December 17, 1961. He was promoted through the ranks, and attained the rank of Deputy Fire Chief. He retired as such in 1991.

- Earl E. Archer was appointed to the D.C. Fire Department on March 1, 1964. He was promoted through the ranks, and attained the rank of Deputy Fire Chief. He retired as such on December 2, 1985.

- Ray Alfred was appointed to the D.C. Fire Department on January 6, 1963. He was promoted through the ranks, and attained the position of Fire Chief. He was selected as my replacement upon my retirement in 1988. He retired as such June 26, 1993. Subsequently, he applied for a fire chief position in the state of Florida, and was selected to fill the position.

- Floyd A. Madison was appointed to the D.C. Fire Department on March 23, 1969. He was promoted through the ranks and attained the rank of Assistant Fire Chief. He retired as such in 1997. He applied for a position as chief in the New York State Fire Department. He was selected as the Chief in June 1999 and served until September 8, 2008, when he retired. Subsequently, he applied for the New York State Fire Administrator position. He was selected and appointed to the position by the State Governor on October 1, 2008, and currently serves as such.

- Donald Edwards was appointed to the D.C. Fire Department on April 19, 1969. He was promoted through the ranks, and attained the rank of Fire Chief on July 27, 1997. He retired as such on November 30, 1999.

- Carlton E. Ford was appointed to the D.C. Fire Department on June 6, 1972. He was promoted through the ranks, and attained the rank of Deputy Fire Chief – AFC-O on January 18, 1998. He retired as such on January 30, 2000.

These men achieved their ranks through education, familiarizing themselves with policies and procedures of the Fire Department, dedication, loyalty, and pride in being firemen. They made me proud in seeing their achievements while in the department, as well as outside the department in similar like service. This would not have been possible in any department, if the Fire Chief had failed to fight honestly and sincerely in the interest of all of his firemen. While this was done in the District of Columbia, I wonder about the Fire Departments in other states.

(See Appendix C for a historical and theoretical view of a no confidence vote.)

Chapter 20

Protests to the Promotion Issue

The Black Fire Chiefs Resource Book, dated September 1, 1985, identified Black Fire Chiefs and showed personnel makeup of Fire Departments in a number of cities in the various states of America. Although the Fire Chiefs were mostly black in all of the departments listed, the rank and file of subordinate staff were drastically disproportionate in all areas. For example, a Fire Department in Georgia had 1 black male Fire Chief, 2 Assistant Fire Chiefs,(2 white), 3 Deputy Chiefs (3 white), 11 Battalion Chiefs (11 white), 72 Captains (70 white, 1 black, 1 Hispanic), 57 Lieutenants (57 white), 92 Drivers-Paid (88 white, 4 black), 6 Inspectors-Fire Prevention (3 white, 3 black), 141 Firefighters (91 white, 48 black, 2 white females). Little is known about the time limits for taking exams in qualifying for promotion, in Fire Departments throughout the country. The Fire Chiefs may very well have been qualified to command the departments; however, it is difficult to imagine any elevation in rank by black firefighters, based on the vast number difference of white staff overall. Having said that, it is not difficult to see what the reaction would have been, or was in some instances, when changes to the status quo were espoused, or direct action was taken. The worst type of behavior in people would surely surface.

The Resource Book reflected the employment ratio between the employees from a racial perspective, with the city population make up being considered. With the Chief being black in all the cities that were listed, the rank and file employees reflected a tremendous disparity in most instances. The Black Chief may have been helpless in reflecting some form of racial transparency. The person most likely to succeed the Black Chief would be white. In some states the department may have been run by white upper management personnel. This assessment is not meant to demean the Black Chief in any way. However, if a recruitment effort is not directed at minorities, then biasness will always be in place, and always practiced.

In the District of Columbia, the percentage of black firefighters on the force compared to city population was considerably out of balance. The overwhelming number of whites eligible for promotion precluded any sizable number of blacks being promoted for a considerable time to come, perhaps, generations. I believe it was necessary to promote the equally qualified black men to the management ranks. That would provide the imagery to whites in the department that black men can be effective leaders in every respect. Also, I believe it was necessary as far as race was concerned, because there were men in the department that served in a segregated military; therefore, their being in an authoritative position most likely caused black firemen to never be promoted to higher positions. Another factor to this conditioning is such that the geography would have a significant impact on any type of personal relationships in being familiar with each other, because whites lived outside of the city limits of the District of Columbia. This also had an impact on the internal black potential recruit to become a fireman. The selective pool of blacks in the District greatly diminished over time, due to a limited human resource.

The union rank and file members entertained the belief that I in some way arbitrarily selected officers to fill positions. They seemed not to acknowledge that all candidates had to appear before a board of inquiry, before any name could appear on a most qualified list. The eligibility list was forwarded to the Chief, who in turn made a selection based on a location by which the vacancy existed, or was anticipated. I was confident that the minorities selected for promotion would perform their leadership roles in a manner that the department would be proud of. Because they had many years of professional experience, harmony was an objective in the performance of the required duty. Although some whites may have been fearful that gross injustices against blacks would somehow be revenged, I don't believe it was seen by the union rank and file. They were against any form of racial harmony in the workplace. However, some members lived outside of the District of Columbia in neighborhoods void of black neighbors; or maybe there was one. It stands to reason what impact affirmative action would have. They had to get over this, and the only way they could do it was with the person himself. Be it black or white, we all have feelings about our races, good and bad; then get rid of those feelings. However, you can't get rid of them until you recognize them. There's not a white man in this country who can say, "I never benefitted by being white." Maybe he doesn't know it. Look at all the privileges you have, too numerous to mention.

One of our country's presidential administrations caused some progress made in Civil Rights to reverse in action. It became known as one of the worst Presidential Administrations on Civil Rights. The tragic and indelible fact is that discrimination against blacks and other racial minorities in this country has perverted our nation's history and continue to scar our society.

During the past 200 years of the Constitution, interpretation by our legal scholars did not prohibit the most ingenious and pervasive forms of discrimination against the black man. From the very point of birth to death, the impact of the past is reflected in the still disfavored position of the black man. The history of discrimination and it's devastating impact on the lives of African Americans, bringing them into the mainstream of American life, should be an interest of the highest order.

As for the white Chiefs, it is reasonable to understand the stepping in a general course of action in how the firefighting service had developed around immigrants from Western Europe, who engaged in services such as police and firemen. These service type jobs afforded them a sense of security, both personal and financial, and a way to fit into a particular community. Also, the men developed a particular kind of competitiveness among themselves through exhaustive physical punishment in the way they handled any dangerous task. This kind of process continually strengthened the making of and belief in the nuclear family concept, which enabled their developing families to achieve greater benefits that life holds. This type of future outlook was solidified by the way a man did his assignment, especially in the Fire Department. This was in conjunction with the satisfactory bonding in a particular engine company or companies that make up the Fire Department. A new recruit had to take his lumps, if you will, to become experienced by demonstrating aggressiveness in becoming proficient in his assignment. Other type of skills were necessary in learning, such as wood workings, mechanical, pipefitting, pottery designing, etc.

This enabled the man to fit into the ethnic equation of the white firefighter, to have something in common, aside from the norm, and become immersed in the developing nuclear family. As this camaraderie was seen by the younger sect, they developed a burning desire to be a firefighter. Today, that sect might inquire about the pay one gets, contrary to yesterday, when men thought of it being an honor, as well as a specialty.

In recent years, the Fire Department has changed in so many ways; for example, my wife Uvaghn was content with staying home to care for our children on a full time basis. This enabled me to feel comfortable in performing my duty as required, while ensuring my family's needs were satisfied. This was a way of life in the white firefighters' immediate family. Also, this enabled him to become a part of a puzzle in the privileged traditional way of life, of many generations gone by. This was a cultural value system as well. Unlike the white firefighters, black men went into the occupation as an opportunity to secure themselves, being of an ethnic group of men, supposedly not having that characteristic of caring for his family. They sought to learn as much of the firefighting technique as possible, and often times exceeded that of his white counterpart. Although black men have been in the firefighting occupation here in Washington, District of Columbia, the Nation's Capital, as well as other cities and towns in our nation.

Granted, this was a welcoming opportunity to be able to prove himself in confidence, dedication, dependability, and qualification in a hazardous occupation. The black firefighter was efficient and effective in his performance of duty. Even though his life style in his segregated part of town differed from his white counterpart, he was no less a dedicated employee in the Fire Department. His verbalism on issues and other subjects of interest indicated that he was not wedded to the Fire Department. The segregated department gave cause to not sleep, eat, and engulf oneself; no, just an employment with a tremendous responsibility, developing a culture of its own.

The cultural difference was illuminated when the department began to integrate. It challenged the wedge, and brought about verbal, as well as physical clashes. You can imagine the hostility that was illuminated when decisions were made in assigning senior black personnel in charge of sections involving white personnel, who were negatively resentful. This action brought about a particular realignment of duties, that in some ways, became an appeasement in what could have been a violent race confrontation. This behavior did in fact happen in other type situations. Black firefighters quite readily recognized a dilemma that they faced in a physical situation of being out numbered. Therefore, they performed their duties in ways that little or no criticism came from superiors, or rank and file colleagues.

There were other acts of antagonism that may have been orchestrated by the refusal of whites to eat together with blacks in the firehouse, sleep in the same areas, or eat with the same utensils. To ensure this, the utensils were broken up by whites and discarded in the trash receptacle. There were many other incidents having racial overtones that could have provided a physical confrontation of a serious matter.

The black firefighters who were the victims of the white rage sensed a kind of mental rigidness necessary to abate their actions. There is an old Negro spiritual that reads, "Trouble don't last always." I believe that our sustaining spiritual faith in God and the church family teachings of the commandment, "Do unto others as you would have others do unto you," are the cornerstone of our survival, in peacefully integrating into the Fire Department, here in the Nation's Capital.

As for civil rights issues, the involvement of an earlier statesman, in particular the abolitionist known as Frederick Douglass, is historically documented. He witnessed various presidents of these United States slowly retreat from the Republican principals of President Abraham Lincoln.

In retrospect there were those persons in southern power during the 1890's that deeply resented the loss of the civil war. The trials of the reconstruction era and the push by former slaves to secure their full freedom rights such as voting privileges and political participation were totally unexceptionable to the controlling hierarchy.

Among the acts of retrogression is one that stands out significantly. It was a well planned insurrection by a specific group of white men in Wilmington, North Carolina, in 1898. A commission study concluded that the violence, which was called a race riot, resulting in the death of an unknown number of black people, was part of a statewide effort to put a specific group of white Democrats in office and stem the political advance of black citizens. This incident is the only known violent overthrow of a government in U.S. history. Afterwards, this group of specific white men in the state offices passed laws that disenfranchised black American people, until the civil rights movement gave rise to the Voting Rights Act in the 1960's, led by the persistence of Reverend Martin Luther King, Jr. and his many millions of followers of all races, creed, religious belief, and political affiliation and overall ideology. However, the issue of voting and civil rights issues concerning African Americans continue.

Voting rights are essential for all citizens to participate in order to sustain the integrity of the Democratic governing process of the American way. When this is denied to any group of citizens, it makes for a violent implosion within a system. For example, Fannie Lou Hammer was a tenacious fighter for inclusion in the Democratic process. When her delegation was denied a seat in the hall of the National Democratic Convention, she led them out of the building. She had been arrested and jailed numerous times. While in jail, she was severely beaten into unconsciousness on many occasions. In spite of this, she continued to participate in the movement toward achieving all rights and privileges due African Americans.

There was a military man who was a West Point graduate. He served in the United States Army as Commanding General of Allied Expediential Forces, as well as the Atlantic Pact Forces. Also, he was chairman of the Joint Chief of Military Staff. His ultimate military rank was Five Star General. Subsequently, he was elected to the presidency of these United States of America. Here was a man that was a leader of both men and women while engaged in war, when people were killed or severely wounded. He saw the worst of the worst conditions that involved mankind. As a leader, he knew what it took to accomplish an objective such as teamwork, courage, bravery, fearlessness, etc. I am sure he realized that it took a complete unification of all personnel in this effort. Our nation's freedom rested on his shoulders, as well as leaders like him, to do likewise to resist an enemy who was hell bent on destroying the allied forces, and their stake in the war.

As president of these United States of America, Dwight D. Eisenhower realized that the survival of this nation depended on the cohesion of its citizenry. He understood the discrimination of the African American citizens in this country, and that he was in a better position to address the situation overall, unlike the limited power to do so within the military structure of which he served.

President Eisenhower was persistent in an effort to guarantee the voting and civil rights for the citizens of the District of Columbia, in spite of the lack of interest by former presidents. He and his wife Mamie refused to eat in establishments that refused service to people of color. They were forceful in bringing about the desegregation of movie theaters and eating places in the District of Columbia. In his order to the Attorney General, a brief in the 1953 Supreme Court case recognized the validity and ordered the enforcement of the "lost" anti-discrimination laws enacted by the Legislative Assembly of the District of Columbia, during the tenure of a Republican Governor.

The favorable ruling thus opened the doors of restaurants, theaters, and other public places to all residents and visitors. He was successful in the passing of legislation for District of Columbia residents to vote for president of the United States, but disappointed that home rule and representation in the United States Congress were not achieved during his presidency.

It is noteworthy that his point man in the United States Congress, Senator Prescott Bush (Republican-Connecticut), among some others, were vigilant in getting the necessary votes for passage of the 23rd amendment, granting the District of Columbia residents rights in the electoral college. The case was made by the Republican leaders that residents in the District of Columbia should have representation in Congress, which taxes them, authorizes wars in which they fight and die, and subject them to all the laws of the United States of America. It was undoubtedly a civic duty as well as a moral responsibility for them to act as they did while in office. These incidents, situations, and references to civil rights issues have its impact in similarities to the practices in the District of Columbia Fire Department, over the many years of its existence. The overall selfish makeup of how the department operated had a contiguity with laws, practices, and conditions in this society's earlier development.

The reference to the military was to illustrate that unity by all persons involved will inevitably be victorious in its objectives, unlike the incident of insurrection that I mentioned, showing how people independently conspired to subvert the civil rights of citizens and establishing rules of their own. Unfortunately, the rules/laws in this case have a profound impact on the overall functioning of every aspect of these United States of America. Now to fast forward concerning the internal workings of the Union Local. I believe that there are times when unions are used by our highest political offices, like gamecocks fighting in a fenced back yard, to highlight the issues of the common man. Indeed a travesty.

Many union members seemed to get pleasure out of seizing the opportunity of a particular issue to confront management in an effort to gain control through intimidation. If they had spent more of their time being supportive and equally creative in the performance of their duties, both they and the citizens of the city would have benefitted tremendously. For an example, there were several telephones in the firehouses. Firefighters, who were members of the union, would contact other members in other firestations to discuss issues that they were concerned with. These issues were brought up in the union meetings for discussions. Unfortunately, when I became aware of the issue of concern, it was aired in the newspapers. The real valid issues could have been more thoroughly addressed. We would not necessarily have been reading comments in the newspapers saying, "The Fire Chief's performance draws heat." Everything that was not sanctioned by the union drew heat, even when I tried bending a little, rather than taking a hard line of equal resistance.

Another example: In 1982, the Fire Department spent over $50,000 on state-of-the art fire helmets, as recommended by Local 36. One year later, they found another helmet they thought they liked better. In an effort to work with them to find the best safety equipment available, a purchase order was prepared and we spent another $50,000. It is my understanding that there were suggestions among the union members to scrap the new helmet, and go back to the initial helmet that was bought in 1982. So, in the same newspaper article, I drew heat by honoring a request by the union and the uniform board to change the shirts from dark blue to light blue, because they were tired of looking the same as the oil delivery men, gas station attendants, and others. They also complained of looking like or being mistaken for bus drivers or police officers.

In an effort to eliminate concerns in the area of uniform identity, I personally designed a patch to be worn on the shoulder part of their shirts. Heretofore, there was no patch identity whatsoever. I tried doing everything possible to show that management was willing to reason with their concerns, but to no avail.

Another concern of the union was my proposal to go from a five man to a four man engine company. To do so, this would save the system millions of dollars. It had been proven that the number of fire deaths had nothing to do with engine company staffing. When a child's life was lost in a fire in the Northwest quadrant of the city, the union used the death of a child to support their position. As I was being badgered by the Judiciary Committee, a member of my staff was notified that the time of arrival at the fire address was one minute from the time the call was received at the Communication Division. The union did everything in its power to discredit me. But when lives are lost, it is a sad day to air a concern, especially when it is known that the statements made are false.

The fire was reported to the command center as smoke in the 1100 block in Northwest. The firehouse closest to that address was Engine 6, Truck 4, Rescue Squad 1, and Ambulance 661. Because of the information received at the Communications Center, Engine 4, Chief 1 responding. However, after arrival on the scene, a box alarm sounded for additional assistance. If the rescue squad had been in service, it would not have responded predicated on the initial alarm received at the Communications Center.

The president of Progressive Firefighters Association released an article to the newspaper, describing the vote taken by Union Local of the predominately-white D.C. Firefighters Association as nothing more than an attempt to discredit my administration by the rank and file members of the local. He went on to say that the action in no way reflected his association's position regarding my administration, an affiliate of the International Firefighters, representing 41 percent of the Districts Firefighters. Also, he stated that the union tactics were reflective of the things they wanted, and couldn't get. He said the no confidence vote was actually false, and therefore misleading to the press.

He pointed out some facts in the union's leadership allegation of "no confidence:" "Our organization has filed a case in court through the Office of Human Rights, dealing with some issues which both the Mayor and Fire Chief have indicated they are supportive of. We have been fighting these issues for years, and the union came in to fight us, and have by spending over $1,000,000 in this effort. The matters that they are trying to negotiate through the public and media are basic criteria that are contained in the scope of our case presently in the Federal Courts."

The Mayor issued a press statement reaffirming his confidence in my position. He pointed out that I was carrying out his request to bring about racial balance in the Fire Department. He felt that the no confidence vote was politically motivated and said that the union was doing all it could to collectively bargain through the newspapers rather than at the bargaining table.

A well-known and respected columnist wrote an article about the issue in the newspaper that read, "The D.C. Fire Department has been a hot bed of racial animosity since the 1950's, when blacks were relegated to separate station houses. Today it remains a department where it takes two groups to represent the concerns of black and white firefighters. The predominately white Local Union of the International Association of Firefighters, which has opposed each of the four Black Chiefs, makes no bones about wanting Coleman out. Their main problem is the Affirmative Action Plan that he has spearheaded, and that the Regan Administration is fighting in the U.S. District Court, challenging its hiring and promotion aspects as discriminatory against white firefighters. The D.C. Fire Department is 60 percent white, and the majority of firefighters assigned to carry out the Black Chief's orders are white union members.

A hearing was held before a subcommittee on appropriation. The subject of the Fire Department came up concerning the overall problem in the Fire Department. The chairman of the committee asked the City Administrator, "What is the overall problem? There seems to be chaos. The firemen do not like the Chief; other people do not like the Chief. They try to embarrass him when he issues an order. What is going on over there at the Fire Department?" The Administrator answered, "The fire service generally is a conservative service. It is the same in almost any jurisdiction. It is the most conservative of the City Services. You will find it is the most resistant to change. The Fire Department has suffered, as all of us know, because we read it all the time in the "news." It has suffered from the racial tensions between blacks and whites, the changing of the command structure, and the issue of residency. Turmoil that was created by the lawsuit, the continual set of lawsuits in part created, or joined by the main

Justice Department, regarding the attacking of the Affirmative Action Plan on both promotions and hiring grounds that the Supreme Court has thrown out. That divisiveness is reflected in arguing before courts of law. Points of reverse discrimination or discrimination tend to divide an organization. Our Fire Department has been divided by publicity, along racial lines. We want very much to solve that, and get beyond it to focus on the Fire Service."

The chairman asked, "Are these lawsuits now behind us?" The administrator responded, "No, we are appealing on both promotions and hiring. The Court of Appeals has enjoined us in effect from hiring any additional firefighters until the case is resolved by the Supreme Court. So, we will be hiring no more firemen, the same thing for promoting. We will not be promoting any firemen until the courts say, and it has been that way for 2 years." That kind of uncertainty and animosity in a situation like that—no promotions, no hiring, no change with everybody frozen in place—generated some of the animosity that was there.

The District of Columbia Fire Department had numerous growing pains in its racial discrimination practices, which were very much in evidence when I joined the department in the early 1950's, through the 1980's. For many other firefighters and myself, we were able to rightfully achieve the leadership ranking in the department, based on qualification; through testing experience, participation, dedication, loyalty, moral integrity, and longevity.

This success story may not have been possible had the Fire Chief, been successful in the eligibility waiting period for testing. The time requirement for testing for eligibility in going from the rank of Lieutenant to the rank for Captain was one year. The Chief took it upon himself to change the waiting time for testing to two years.

There were nine of us (all black), who were eligible to take the examination. I was able to get the other eight Lieutenants to accompany me to a meeting with the Chief to ascertain when and why the change was made. Our meeting with the Chief was unsuccessful for reasons that he would say that the change was the way he wanted it.

We all had studied hard for the exam, and if this change was valid, it would have been devastating to our self motivation. In view of the existing conditions, as well as the time frame, I suggested to the others involved that we consult legal counsel, and they agreed. The results were such that the City Council had not issued a special order; therefore, the Fire Chief had no formal authority to otherwise, or arbitrarily, change policy. As a result, the nine of us were able to take the examination on schedule, resulting in four of us being promoted to Captain. The practice of arbitrary changes in all forms and procedures within the Fire Department, in all levels of leadership reflects a systemic form of reluctance.

Since the challenge to the unofficial proclamation was successfully overturned, many firemen who benefited from this, credit me for their having been able to climb the promotion ladder. Many of the men achieved the ranks of Captain, Battalion Chief, Deputy Chief, Assistant Chief, and the ultimate as Fire Chief. I said to them in response, "If you see or hear of something that is not within the rulebook, exercise tact in pursuing an end result to any and all things that's believed questionable." I reminded them of a similar situation that Lieutenant Ralph Smith encountered when he was eligible to take the examination for the rank of Captain in 1945.

These type of practices clearly showed a complete disregard for an employee's well being for himself and his family, a disregard for the morale in the work place, a disregard for the safety of the citizens that the department is obligated to serve, as well as protecting the city infrastructure. The effort to impede the potential creative development of all department employees was viewed as reprehensible.

Chapter 21

Standing on the Shoulders of Others

In my career as a firefighter in the District of Columbia Fire Department, and retiring as the Fire Chief of the Department, I stood on the shoulders of so many whom I had admired. I chose to emulate their courage, loyalty, character, and perseverance. Some whom I hold in great esteem were those in the United States Military Service, in particular World War II. I refer to the war, because I too was a member of the regimented Armed Forces, as were many of the firefighters in the District of Columbia. Many of these men demonstrated the kind of comportment that was evident of training. As a result, many have demonstrated their loyalty, dedication, respect for authority, and perseverance in their performance of duty obligation.

I stood on the shoulders of many African Americans who paved the way, and paid the ultimate physical price, that persons like myself could possibly achieve a better way of life for ourselves and families. Having said that, my remissness would be unforgiveable if I failed to acknowledge a friend and colleague, namely, Fire Chief Burton W. Johnson, the first African American, to achieve this title.

Chief Burton W. Johnson joined the D.C. Fire Department in 1943 as a recruit. Subsequently, he was called to serve in the U.S. Army. He became a First Sergeant, was discharged, and returned to the Fire Department. He was assigned to Engine 4, the all colored station, located at 931 R Street, N.W. He reenlisted in the Army and was honorably discharged as a Master Sergeant (E7) Infantry in 1946. He returned to duty at Fire Engine 4.

Chief Johnson served under Captain Keyes for a considerable period of time as directed. He was promoted from private to the rank of Fire Inspector in 1956. He was the first of his race to achieve the following ranks:

1. Sergeant in the Fire Prevention Division

2. Lieutenant in the Fire Investigation Section, Fire Prevention Division

3. Captain, Commander of Fire Investigation Section, Fire Prevention Division

4. Fire Marshall

5. One of the first Battalion Chiefs

6. The ultimate position of Fire Chief here in the Capital of Washington, District of Columbia, in 1973.

Chief Johnson initiated a number of significant projects, and his female appointment to the department was a first. He instituted the modernization of the Emergency Reporting and Communication System, which allowed the department to abolish the "pull type" fire alarm boxes that reduced false alarms by 50 percent. Chief Johnson had two new firehouses constructed under his leadership, Engine 4 on Sherman Avenue, N.W., and Engine 2 at 5th and F Street, N.W. He, along with the local International Association of Black Professional Fire Fighters and the Progressive Firefighters Association, received national and international recognition.

Chief Johnson was a man who was very easy to converse with during my early beginnings as a firefighter. I cannot over emphasize how much information and encouragement he gave to me, while I was being indoctrinated as a firefighter. He was a wagon driver; as a matter of fact, he was the only one who could start his vehicle, because of the method that was necessary to do so. He impressed me so much that my ambition was to be a wagon driver; and I did that.

He was always eager to answer any questions that I asked concerning the rules and regulations practiced in the department. His answers were very precise and easy to understand. However, he would stress that I should always conduct myself respectably, so as to command respect as a firefighter/person.

Chief Johnson left such an inspiring legacy of his accomplishments in the Fire Department. His numerous community and social membership activities spoke volumes of his humility, integrity, and love for mankind.

He was a person that I chose to emulate in his professionalism as a fireman. He made all of us proud.

The Black Fire Chiefs Resource Book that I mentioned earlier paid homage to a Fire Chief by the name of James H. Shern of Pasadena, Calif. He retired in 1981 and expired in 1982 due to health reasons. Here is an excerpt as it relates to him and his achievements: Past President of the International Association of Fire Chiefs (IAFC), and the former Chief in Pasadena, Calif. "Chief Shern was one of the most respected and admired in American Fire Service," says the president of the Los Angeles County Chapter of the Chiefs Association. The current president of IAFC, of Tulsa Oklahoma, said the fireservice and the association would miss Chief Shern as a friend and a professional. The general manager of the IAFC praised Shern's leadership. "Chief Shern demonstrated remarkable insight into problems of the fireservice, both on the national and local levels." Chief Shern was a pioneer in the area of minority representation in the fireservice. He was the first Black Battalion Chief in the history of the Los Angeles City Fire Department and the first Black Fire Chief in a city of more than 100,000 population.

In 1972, the Pasadena Fire Department had 2 black firefighters out of a force of 167. Of the departments 128 firefighters, 21 are black, 4 are Hispanics, and 1 American Indian.

Chief Shern was known throughout the country as one of the most knowledgeable people in the field of Fire Protection. He has been published in almost every fireservice periodical and held membership in almost every national fireservice organization and many regional fireservice organizations in the Western United States. Chief Shern's research contained in Smoke Towers in Central Core Structures, published by the National Fire Protection Association in 1965, is now a standard code for many cities. He served on the Boards of Directors of the Boys Club and St. Luke Hospital in Pasadena, Calif., the Scripps Home in Altadena, and was an honorary member of the Pasadena Chapter of the American Red Cross.

Chief Johnson and Chief Shern persevered in their obligation to duty, in spite of the racial practices/discrimination in the Fire Departments. However, through their accomplishments, they proved themselves to be men who saw the true meaning of being a fireman, in saving lives and property in our towns and cities.

The Black Fire Chief Resource Book included a poem dedicated to John B. Stewart, Jr. on his promotion to Lieutenant in March 1966. It was later rededicated to all black firefighters.

Tribute to a Black Firefighter

I heard the engines clanging gongs a block or two away. And when I saw
the raging fire, dark smokes and waters spray.

I saw the shinning ladder, as it reached up to the wall. And then I saw
him climbing, climbing upward, toward the call.

His black hands gripped the ladder, which he climbed with sured pace.
the smoke engulfed his body, flames danced about his face.

"I can't hold on! Please help me!"A youthful voice, a pleading cry.
"Hold on! I'm coming!" was his firm assured reply.

The roof began to crumble. The building end was near. Those below
began to scatter at the sound, which filled their ears.

His dark face was gripped with horror, his mind was seized by fear. As
he reached the fiery window. He heard, "Swing the ladder clear!"

In that next heroic moment as I closed my eyes to pray, a black hand
grasped the child and lifted him away.

There atop the ladder clearly seen by every eye, were the fireman and
the child dark silhouettes against the sky.

He was grimy, hot, and haggard. As he stepped down to the ground,
a cheer arose "No name please!" "Compared to bigotry and other
barriers I've overcome, this was an easy day."

James O. Rogers, Author
Blues and Ballads of a Black Yankee

Dedication of Engine Co. 4
In Memory of the late Fire Chief

Burton W. Johnson

Chapter 22

Magazine Publication

A white lady came to my office one day seeking to interview me concerning the overall operation of the District of Columbia Fire Department, in practically every detail. Her name escapes me at this time; however, prior to any discussion, she assured me that I would be given a copy of her writings for my personal perusal and approval prior to any publication. She presented identification reflecting her representing the magazine. She asked permission to visit some of the firehouses, take photos, and interview firefighters of selection. I asked her to follow me to the firefighter training facility located in the Southwest sector of the city, for a firsthand look at some of the techniques involved in firefighting. She asked to take a photo of me with the firefighting men. She was given a timely cursory of as many areas in the department of her choosing, after which, she departed.

It was well over a month, and I had not heard anything from a representative from the magazine. However, well into the 2-month time frame of no news, I received a telephone call from Deputy Fire Chief who is black, incidentally. He asked me if I had seen a draft of the article written from their interview of me and others in the department. I told him no. He said that he would bring a copy for me to see, and that I would not like what the contents were. The Deputy Chief and I sat in my office, and discussed some of the contents within the draft. I agreed with his statement that I would not like what was in it.

At no time did I ask the Deputy Chief how, where, or when he received the article's draft copy. I felt that if he wanted to tell me, he would have. He and I had been, and continued to be, longtime friends. As a matter of fact, he and I were very competitive; along with a few others, we studied rather intensely for the rank of Sergeant examination. We both made the eligibility list, and he scored higher than I. However, we were both promoted at the same time. The Deputy Chief and I challenged each other in seeing which one of us would score on the next level promotions. I scored higher than he on all other promotions. I left him at the rank of Deputy Fire Chief. He and I continued our cordial friendship.

As for the draft, I did not call anyone to discuss any of its contents that I considered questionable. Again, I did not question the Deputy Chief as to how he was able to obtain a copy of the draft. However, it would not have made any difference, because the magazine was already being distributed, and I received a copy. I became increasingly vexed the more I read. The article was replete with so many untruths and opinions that were unsubstantiated and without any merit. After reading the article several times, I was not clear as to the author's overall objective. The article was so convoluted with distortion that I chose not to comment any further on any content in the article that was so distensible in distorting the character of our second to none Fire Department, here in the District of Columbia.

The article was written by the author, and characterized in the cover story as "The burning issue at the Fire Department is Chief Coleman himself (under fire), (trouble)."

The article grossly tarnished the image of the dedicated hard working and otherwise faithful firefighters serving under my leadership. The article delved into the racial makeup of the department, as to how each race see things differently. "White firefighters claim that Coleman tries to harass and intimidate them, and that many of his orders are racially motivated. As a result, long dormant racial tensions have flared in the city's firehouses; fistfights between blacks and whites have become commonplace. Firefighters charge that Coleman has practiced reverse discrimination in making promotions, jumping blacks ahead of seemingly more qualified whites. As a result, experienced mid-level officers have quit the department in disgust, and other firefighters have lost confidence in their supervisors." The article presented some distinct analogies; i.e., it states that the Fire Department has a long history of racial discrimination, and that it was born during the nation's first great struggle for racial equality, the Civil War. The department acquired an esprit de corps that was unique among the District Agencies.

Firefighters really loved their work. Most of them were working class young men from Maryland and Virginia suburbs. They held down two jobs. It was not unusual for firefighters to follow in their father's footsteps into the department, and for their sons to join up in turn. They arrived at the station, hours ahead of schedule, cooked large meals together, and ate. However, a published article states that in 1919, a decision was made to place all black firemen in one engine company in a neighborhood of largely black population, and the department remained segregated until the early 1960's.

The District's black population grew steadily, yet the Fire Department remained a bastion of Jim Crow, due to low turnover, and policies that discouraged blacks from applying. By 1968, only 15 percent of the firefighters were black. The white dominance remained. One of many glaring examples was the treatment of Fireman Smith. He was the first black in the department to rise to the rank of Battalion Chief. It was understandably clear that he endured many hardships, and lawsuits were necessary to abate the harsh treatment by the white rank and file firefighters. The wrath of these men were unleashed when firemen such as he would show intentions of wanting to advance their status in rank. Battalion Chief Smith joined the department in 1932 and was assigned to a black only engine company (No. 4) located on Virginia Avenue, Southeast. He scored well on his exam for Sergeant, but was unable to advance in the company due to three black officers occupying the only available positions.

During the war in 1945, many black men joined the department, causing a new station to be opened. Officer Smith was promoted to Lieutenant, and assigned to the new unit. He was denied the opportunity to take the test for Captain. The establishment decided to stymie his chances by changing the waiting time to 5 years; this was revenge punishment. In 1950, when the 5-year waiting period was up, they changed the rules again. This time the wait was 20 years before he could take the exam for Captain. He took the exam and placed 5th out of 44 examinees. He was promoted in 1958 at a time when the Federal Government was applying pressure for the department to integrate. Smith and other firefighters reflected on some human practices before integration, such as having to sleep in beds restricted to blacks, as well as using respirators restricted to blacks. The whites destroyed the dishes that were used by blacks the previous night.

Smith sought the position of Battalion Chief in 1965, he was snubbed in favor of a less experienced white Captain. He appealed to the D.C. Human Rights Commission, who determined that white officers lied about his abilities in order to block his promotion. A sympathetic white commissioner went to the then Fire Chief, and said, "You don't have to promote Smith, but not another son of a bitch is going to be promoted until he is." The Chief relented and his revenge was "We're going to get this nigger" by assigning Smith to become guardian of the department's soap and blankets. Smith was fed up at this juncture and retired several months later. Many firefighters give him personal credit for his extraordinary efforts in forcing changes to be made in the D.C. Fire Department. While he accepted this recognition, he admonished the black firemen to study the books to become more proficient in their duties and responsibilities.

The president of the Progressive Firefighters Association, a black organization, said that black officer statistics were still disproportionately low in numbers, and wholeheartedly supported my efforts to abandon the traditional promotion procedures in order to produce more black officers. As Chief, I said the old-boy syndrome is nothing but discrimination. There is no way that I could be a good old boy to everybody.

While I have mentioned the tenaciousness on the part of my fellow firefighter, Officer Smith, there was another outstanding individual African American, namely, Garrett Morgan, who, like so many other men and women that I have acknowledged in my writing, went about inventing numerous devices that has had an enduring impact on not only citizens of the United States of America, but the entire world. He, like Officer Smith, was seen as a man on a mission.

Garrett Morgan and Officer Smith were similar in their pursuits. Garrett Morgan was granted a U.S. Patent for inventing the first traffic road signal. The patent was granted in 1923. The traffic signals were also patented in Canada and Great Britain. One of his many inventions was profound in securing the safety of a firefighter, in the process of saving the lives of others; and that was the invention of the safety hood and smoke protector, that allows firemen to breathe freely for a considerable period of time, withstanding suffocating gasses and smoke. The invention was patented in 1912. A later model gas mask was patented in his name in 1914. He became widely recognized when he and a team of volunteers saved 32 men that were trapped in a toxic gas filled water works tunnel, 200 feet beneath Lake Erie. After the rescue, he won gold medals at the second International Exposition of Safety and Sanitation, and from the International Association of Fire Chiefs. The inventor received orders from Fire Departments all over the country. Unfortunately, many of the orders were cancelled when it was discovered that Garrett Morgan was black. Sadly, many chose the face of danger and death, rather than buy a life saving device created by a black man. However, business boomed when Morgan hired an actor to pose as an inventor, while he dressed as an Indian. The actor would announce that Big Chief Mason "would go inside a smoke filled tent for 10 minutes, and emerge unharmed after 25 minutes." The onlookers were simply amazed. Morgan's life ended in 1963. He left a marvelous and lasting legacy in American history.

The magazine article stated that the Fire Department, for decades, was the "whitest" agency. The article referred to the 1960's as a time when black firemen had limited opportunities for promotion, and blacks at the top were disproportionally low. Only 20 of the department's 70 Sergeants were black; of the 34 Lieutenants, only 10 were black; of the 64 Captains 7 were black; of the 36 Battalion Chiefs, 4 were black. It went on to say that the Mayor had charged me with redressing the department's racial imbalance. The article pointed out the fact that therein lied the root of the problem, putting me in the thick of what had become a common American problem. It goes without saying, any attempt to right the wrongs of discriminatory practices of yesterday, and unfortunately, still today, causes a tremendous morale problem, as well as physical conflict at times.

A picture showed me surrounded by a host of firemen, holding an ax, extended blade in front of me. It seemed to portray me as " using an ax when a scalpel is required." It went on to say that a union representing the firefighters jammed the tiny union hall to cast a no confidence vote in my leadership, by a tally of 190 to 2, that I should resign. Some of the reasons given were that I was practicing reverse discrimination in making promotion of blacks ahead of seemingly more qualified whites, and that I intimidate whites in my orders that were racially motivated. A black firefighter was quoted as saying that I put out orders that were intended to create tension, and that some of the kids in his boy scout troop could run the department better.

Some other comments were of a negative assessment of my leadership: "We have gone a long way backwards in racial relations," said a Captain, at Engine 11. A former Assistant Chief said, "They think you are trying to pick on them, because they are black." "I am worried about the force," said a retiring Battalion Chief. A Sergeant at Truck 13 said, "We are not opposed to affirmative action, but we don't believe that the definition of affirmation is giving someone something he does not deserve." A firefighter at Engine 6 said, "It's amazing we were still operating." A Sergeant at Engine 13 said, "Coleman has been the worst thing that's happened in the Fire Department that anyone can remember." A fireman at Engine 6 said, "If Coleman was working for a private company, he would lose his job in a minute." A Sergeant at Engine 31 said, "The way the department is being run, you don't know what the rules are anymore, they're changing every day." A firefighter at Truck 16 said, "Coleman has kind of forgotten where he came from, he was down in the trenches once; now that he is the boss, he doesn't have to prove it to anybody." Those persons who were interviewed were members of the union, who opposed nearly everything that I proposed, as well as firemen who were disgruntled for various reasons. The President of the Progressive Firefighters Association, said, "Even though we have got a black Fire Chief, the power structure is still massively against us."

The article further stated that despite the departments changing demographics, many of its blacks could not forget the bitter years of white dominance. I was quoted as saying, "I don't think the problem that exist have any bearing on my ability to get the job done." I believe that to be a true statement. My objective was to manage the District of Columbia Fire Department to be among the best in the nation.

The publisher seemed to cast disparaging comments on my management style of the Fire Department, also the Mayor's management was a cancer on the city, and that a congressional investigation should have been conducted. These remarks were made to the downtown Franklin Square Associates. His remarks were worded bluntly at the event celebrating the revitalization of the downtown business area. While he continued to berate the management styles of running the Fire Department, and the city overall, it was understood that several council members agreed with his criticism.

The Mayor said that in his many years in office the publisher had never been in city meetings with him and had never asked to bring his editors for interviews of how the city functions, and as a journalist, owed it to his readers to at least look at what the city is about, and that his remarks were of a personal vendetta.

In reviewing some of the many statements that were made about my management style, especially the most negative ones, I saw those persons as feeling insecure in their abilities to do what it took to excel in efforts of becoming a professional firefighter, and to envision the possibility of being the Fire Chief. In others, I recognized a form of misdirected anger of what they heard, and were convinced to believe. As for the black firefighter, who commented by saying that I had forgotten where I came from, if he is comfortable in staying in the trenches, he should become adapted in whatever direction he is assigned to follow by upper management. I say to the statement he made about me not remembering where I came from, walk a mile in my shoes of the past; not only mine, but in many of his ancestors' shoes, especially those who functioned in uniform, while fighting in the American Revolutionary War as slaves. They had no reason to fight; however, they had the good common sense that their lives depended on how to follow orders in concurring an objective. This was a segregated fighting force in 1775. The practice should have ended in 1776, but it took approximately 170 years for this military type of segregation to end. The courageous President of the United States, who signed the executive order, was Harry S. Truman. This action allowed men and women of color to achieve all ranks in the service to Four Star Generals and Admirals, also to soar into the universe as astronauts.

As for me and my management style, I feel that I had to employ many hours of undivided attention to the details of being in the department at all levels of responsibility, especially that of Fire Chief, and surmount whatever zealousness it took to succeed.

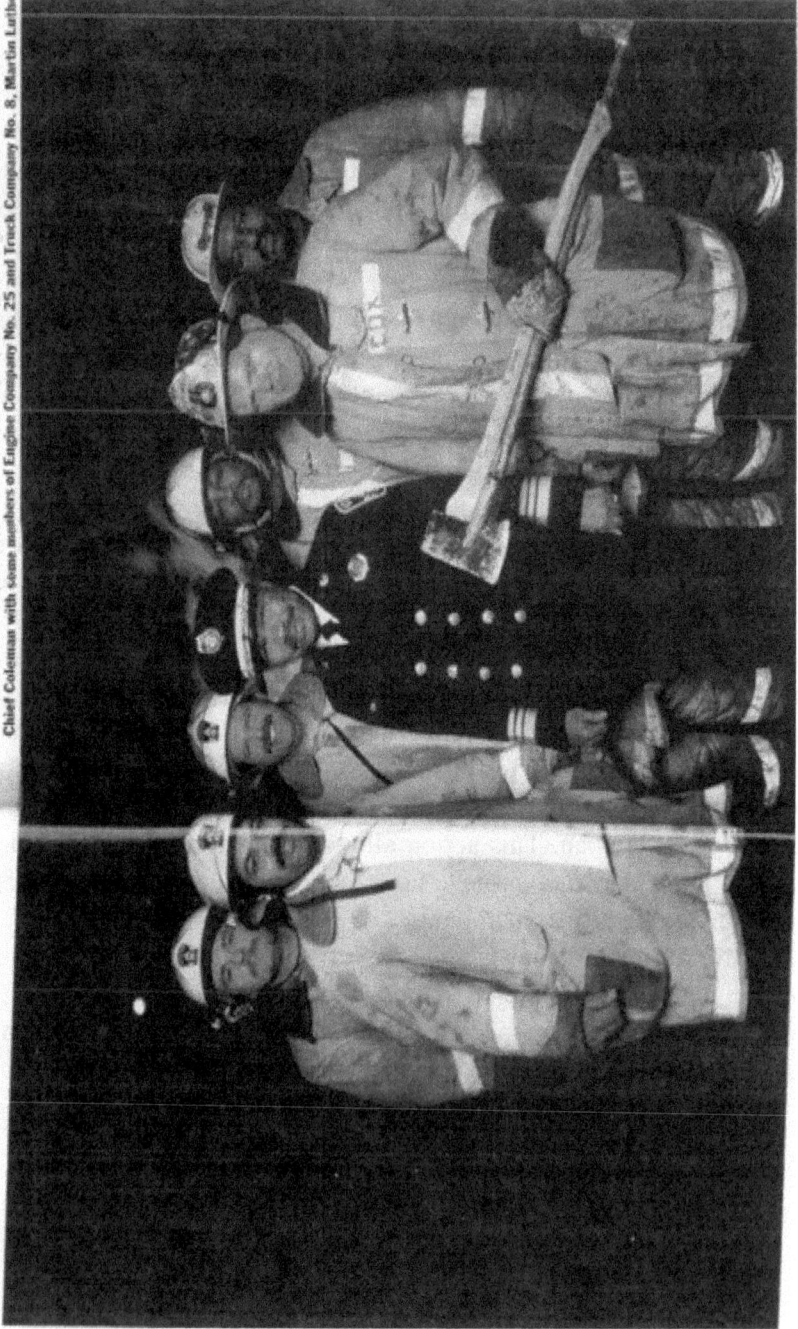

Fire Fighter, a 25-year veteran, Coleman. The image by calling Engine Company. Also a ___ Ladder

Chief Coleman with some members of Engine Company No. 25 and Truck Company No. 8, Martin Luther

Chapter 23

The Fireman's Parade
August 1988

Upon receiving the information from the International Association of Fire Chiefs (IAFC) Board of Directors that our city of Washington, D.C., had been selected to host the 115 annual conference, I began appointing a select number of chief personnel to establish a plan to accommodate the convention attendees with the customary parade, at the conclusion.

Chief James Tate selected the principal officers to spearhead the project. Chief James Tate was designated to develop all aspects of a parade, since he had done this every year since 1982. I felt there was a need for a firefighter's parade to be of multiple purposes.

Chief Tate presented a plan that would demonstrate to the citizens of Washington, D.C., the firefighting equipment that their tax dollars pay for, in order to protect their lives and property, and give them a chance to see the complex make-up of these machines. It would reflect the training of our dedicated firefighters in the use and maintenance of the equipment and show the public how far the apparatus had changed over the years, and how new technology was enhancing the effectiveness in the use of the equipment. The display of firefighters and their equipment would be an attraction for young men and women to join the Fire Department in some capacity.

When word of Chief Tate having been selected to oversee the parade development got around, he was called immediately by other ranking colleagues who expressed their willingness to assist in any way he wished them in this endeavor. I was sitting behind my desk when the telephone call came in from Deputy Chief Tate informing me of the persons who had called him. When I heard the names, I stood up while listening and said to myself, "With these dedicated firemen who put their lives on line daily to rescue and save the lives of others, I know that they will ensure that this parade would make the City Government, and its citizens proud." The officers/firefighters and office personnel offered to assist Battalion Chief James Tate in the parade development.

At this juncture, I believe it appropriate to mention some unpleasant and humiliating conditions Chief Tate experienced in the Fire Department. He was assigned to the firefighting unit at Engine 29. Early on in this assignment, he encountered some resentment of his presence, in that he was a black fireman under predominantly white officer supervision. His work ethic was without blemish. He was a quite type individual who was a perfectionist in his performance of duty. Chief Tate was transferred into various departments upon being promoted, such as a Captain in charge of the personal equipment of firefighters. In this particular area, the Deputy Fire Chief was white. Whenever the Chief had reason to be absent, he would designate a Sergeant to hold the keys to the property rooms. The Sergeant was also white. Captain Tate could not effectively perform his duties of responsibility without having complete access to all designated areas. This action in itself caused Captain Tate extreme humiliation in this disrespect for his rank. However, he never wavered in his responsibility under the adverse condition. Captain Tate continued to maintain a professional demeanor, eventually attaining the rank of Battalion Chief, and assigned to the Apparatus Division. In the latter department, he submitted drafts for new equipment as specifics for the forthcoming year. He had his staff to schedule equipment for repairs, and restored as expeditiously as possible. He made certain that all equipment was in operating condition at all times.

Later on in Chief Tate's career, I recognized that he continued to exhibit the quality of a person that could be counted on to complete any major task assigned to him. I asked Chief Tate to develop a type of device to remove soap residue from cars. The reason for this is; I saw a need for smoke detectors to be provided to city residents that could least afford to purchase them. As a result, a plan was devised to make this a reality, so my family, wife Uvaghn, sons, Andre', Theodore, and Michael, and daughters, Sandra and Yvette, washed cars every Saturday during the summer months, weather permitting. Others involved in this endeavor were members of the Administrative and Apparatus Divisions.

This effort was enhanced by the creative ingenuity of Chief Tate. He devised a sprinkler system to rinse the soap from the cars before the final wipe down. The finance realized from this effort allowed for the procurement of sufficient smoke detectors for residents of the city who needed them. Firefighters were dispatched to the homes of residents in order to assist in the proper installation of the detectors. This effort was in keeping with the Fire Prevention Proclamation, signed by the Mayor, District of Columbia. I understand that this was the first of this kind in the nation.

Now to resume the efforts in the development of the parade. Battalion Chief Tate informed me that he needed a support person with the rank of Deputy Chief, to assist him in some situations that could occur. I asked him for a name of a Chief that he felt would be of value to him. His reply was a Chief that's assigned to the Ambulance Division. I told him that I concurred whole-heartedly, and that he made an excellent selection.

Chief Tate and all of his support staff were meticulously prepared for this day of event. He continually apprised me as to the progress being made in the development of the project on a weekly basis, and nearing the two months before the parade, a memo from him was delivered to me in my office daily, and followed up with a telephone call when warranted. Finally, Chief Tate suggested a meeting by all key personnel to inform me of all details concerning the project. At the conclusion of the meeting, it was unanimously agreed that all specifics were in place.

The parade started at 12th and Constitution Avenue, N. W.. It lasted 4 hours, ending in Southwest, Washington, D.C., near the Fort McNair Military War College. The parade was ingratiating by others in seeing so many groups participating, who came from both near and far with their assortment of apparatus. Some equipment was in the antique status; however, exquisitely maintained. There were numerous marching groups representing high schools in our city, as well as others from the outside. Also, there were police groups on horseback and motorcycles, as well as majorette and drum corps, Scottish pipe bands, Fire Department pipe and drums, Police and Fire Department color guards, Ronald McDonald railroad engine replica, etc. The parade was viewed by thousands of spectators. At the conclusion, an area had been readied with trophies for presentation to the deserving.

Due to the success of the parade, Chief Tate and his support staff were given the highest form of recognition possible for their outstanding performance in this endeavor. It was such a success, it made us all really, really proud and appreciative for having represented the citizens of Washington, D.C., in a class style showing.

I personally want to thank Mrs. Josephine Tate, for providing the photos of her late husband, Deputy Chief James Tate, that appear in this book. Hopefully, this will reflect as a tribute to Chief Tate, in being a loyal and dedicated officer, in my cabinet, as well as to the citizens of the District of Columbia.

Car Wash Fundraiser for "Smoke Detector Giveaway Campaign"

On August 29, Fire Chief T.R. Coleman held the fifth in a series of car washes to support the Department's "Smoke Detector Giveaway Campaign."

Chief Coleman said: "Occupants of homes with working smoke detectors double their chances of survival if a fire occurs. Through this campaign I remain committed to my goal of giving every citizen of the District of Columbia the opportunity to better protect their homes and loved ones from fire-related catastrophies."

The D.C. Fire Department has held five car washes this year. Approximately 750 cars have been washed, 500 smoke detectors have been purchased and another 500 ave been ordered.

Fire statistics show the United States has one of the highest fire death rates, per capita, in the world.

Of all fire deaths, about 80 percent occur in the home. During a live on-air radio show with WOL-AM from the car wash, Chief Coleman stated that about 25 percent of the residents in the District of Columbia do not have smoke detectors and are at the highest risk from fire deaths.

The most recent car wash was held at RFK Stadium parking lot, where Fire Chief Coleman and volunteers from the fire department pitched in with gusto. Proceeds went towards the purchase of smoke detectors for any District resident who does not have one and cannot afford to purchase one. **DISTRICT OF COLUMBIA LAW REQUIRES EVERY HOME TO BE EQUIPPED WITH A SMOKE DETECTOR.**

The next car wash is scheduled for September 26th, at the Stadium parking lot. Volunteers are needed. Please contact the Public Affairs Office on 745-2331, with your name and time you will volunteer to wash cars.

D.C. Fire Department Community Services

The District of Columbia Fire Department's mission is to protect life and property against devastating fires through its Fire Suppression Force and Ambulance Service.

In order to educate the public in proper fire safety, protection and prevention behavior, the DCFD provides such community services as:

Home Fire Safety Survey Program

Uniformed firefighters survey accesible homes, with the permission of the occupants, to help residents eradicate fire hazards and educate them in good fire prevention methods. Residents are warned about such hazards as overloaded electrical outlets; non-working smoke detectors; faulty wiring; the placement of combustible materials too close to heating units; and tucking extension cords under carpets.

Institutional Fire Safety Program

This is a collection of seminars and workshops designed to enhance the knowledge of staffs and residents of institutional care facilities with fire safety survival techniques.

Fire-Safe Living Program

This program informs residents of public housing of fire-safe procedures that should be followed on a day-to-day basis.

Smoke Detector Give-a-Way Program

D.C. Law 2-81 requires that all dwellings in the city be equipped with operating smoke detectors. The DCFD encourages all residents to purchase and maintain the detectors in proper working order.

continued on page 2

∴ Marion Barry, Jr.
▬ Mayor

T.R. Coleman
Fire Chief

Message From The Fire Chief

The Senior Executive Staff continuously measures the performance of the Fire Department against that of our competitors from other jurisdictions based on available statistical data. I am pleased to report that every analysis of competitive performance I have seen recently substantiates the soundness and vitality of our Department.

The Fire Department's perfor-.1ance, relative to its fire service and fire prevention programs, is the best evidence available of the quality of our management. I believe that the Fire Department has a truly superior team that performs well in each area of its operations.

One major concern is the ambulance crises. The primary objective is to provide fast and efficient service to the residents of and visitors to this city. The Ambulance Service responds to approximately 120,000 calls a year on a 24-hour basis (with 21 ambulances). These 21 ambulances are required to serve a resident population of 627,400.

The Department is working closely with the medical community and other District agencies to identify ways to improve the ambulance response time. I feel confident that in the very near future, along with our fire operation, this city will be the beneficiary of one of the best mbulance services in this country. Nonetheless, the numerous letters of commendation received recently for ambulance services rendered is a clear indication that the members of the Emergency Ambulance Bureau

and the Fire Fighting Division have sound reason to take great pride in their ability to quickly react to urgent community need, even during the brutal weeks of the ambulance crises.

Our goal is to improve the Ambulance Service to the point of excellence, and I think we are almost there.

The choices we are faced with and make everyday of our lives determine the direction in which we grow, the ways in which we change, or if we grow or change at all. By taking responsibility for our own lives, not blaming others, we will discover the power of forgiveness. Any piece of behavior can be viewed as simply doing or not doing. Failure or somebody elses editorial opinion must not be feared and should not affect our feelings of self-worth.

As your Fire Chief, my management philosophy is simply to do the best job that we can do everyday. That will determine the direction in which we grow.

———

continued from page 1

For those residents who cannot afford to purchase a smoke detector, the DCFD will provide one free of charge. For further information on how to obtain a free smoke detector please contact the Community Relations Unit at 745-2343.

Bicycle Registration

D.C. residents may register their bicycles at any fire house. This procedure will aid in tracing stolen or missing bicycles. Bicycle registration is from 8 a.m. to 8 p.m., 7 days a week, for a $1.00 fee.

Voter Registration

District residents may register to vote at any D.C. fire house.

Sharpe Health School Program

This Fire Safety Education program for handicapped citizens is specifically targeted to reach children with learning disabilities and/or multiple physical handicaps. The program focuses on special building evacuation procedures for this segment of our population.

Junior Fire Marshal Program

The program is designed to reach 3rd and 4th graders with basic information on fire safety, prevention, protection and burn prevention. This instruction is offered in the D.C. Public Schools during the school year.

Smokey The Bear Program

The Smokey The Bear Fire Safety Program is for pre-kindergarten through first grade children. They are taught how to protect themselves and their families from the hazards of burns and fire. "Smokey" also teaches them how to reduce the number of carelessly caused fires at home as well as in forest and park areas. The children get to know who "Smokey The Bear" is and how to follow his rules.

Smokey The Bear Visitation Program

The D.C. Fire Department has received a "Smokey The Bear" costume from the U.S. Department of Agriculture that is used in the public safety awareness program.

Partners in Education Program

A fire safety education program is conducted for junior high school students at Douglass Junior High School, which is the DCFD's partner in education.

"Be Fire and Burn Wise" Alert

A monthly newsletter that reaches approximately 97,000 persons with fire burn safety messages and safety tips from the Fire Chief is distributed to the entire D.C. Public School system.

"Fire Hawk" Juvenile Firesetter Program

A program designed to reduce the risk of injury and loss of property from fires set by children. This program addresses the problem of juvenile fire setters and what the Fire Department and community can do to help these children.

Fire Safety Education Center and Mini Museum

Located at 438 Massachusetts Avenue, N.W., the center and museum house such historic items as a 1905 horse-drawn Steamer Wagon. The museum is open Mondays-Fridays, 9 a.m. to 4 p.m. For tour information call 745-2343.

Chapter 24

The Missing Photo

In July 1982, I went to San Diego, Calif., accompanied by a staff member associated with the Public Relations Agency in the District of Columbia. The purpose was to address the Convention as a representative of the Fire Department in Washington, D.C., and appeal to the convention authorities to hold the International Fire Chief's Annual Conference in the Nation's Capital in 1988. I assured the convention attendees that the District of Columbia Government and Fire Department would be proud to host the conference. My expression to them was come to Washington and experience the hospitality of the area's fire services, see our Nation's Capital, and take home a wealth of ideas to assist their Fire Departments.

The convention's scheduled dates were August 27 thru August 31, 1988. In preparation for the convention, a handbook was printed by a printing company in Chicago, Ill., that had been printing the Fire Chief Magazine since 1970. In each of those prior years, the Fire Chief's picture from the convention's host city was shown on the magazine's front cover. The magazine for the 1988 conference displayed the photo of the Jefferson Memorial on the front cover. Since 1970 and all subsequent years, the front cover had always shown the picture of the HOST CHIEF. However, much to my chagrin, my picture was not on the front cover, nor was it anywhere in the magazine.

The 1988 magazine was printed with red and blue colors throughout. It was replete with photos of fire and rescue operational scenes, fire equipment displays, pictures of the host city, historical landmarks, and in this case, a picture of the pandas in our zoo frolicking with each other. In other parts of the magazine, a detailed description of daily activities were printed, such as meeting times, as well as entertainment facilities, eateries, shopping malls, etc; as well as the location of the International Association of Fire Chief's Headquarters in Washington, D.C. In the area of the book where the Host Chief's welcome comments would be, was the IAFC President's picture, along with his welcoming address:

"Welcome to IAFC 88."

Dear Fellow Fire Service Leaders:

I am particularly honored to welcome you to Washington, D.C., and to the 1988 IAFC Annual Conference. This year's conference will be like no other you've ever experienced.

We have made this year's meeting a metropolitan Washington event, joining the talents of not only the District of Columbia Fire Department, but those of other fire departments in the area.

The program we have put together for you this year surpasses every previous conference, and will keep you on the leading edge. There are numerous educational sessions, including a number of top- notch national and international speakers from government and industry. They will address current issues as well as emerging management practices, and technical trends. Our exhibit area will cover nearly 8 acres, with hundreds of exhibitors on hand to discuss their products and services.

Washington is well recognized as a global crossroad, serving our nation's capital, and providing a forum for international diplomacy and culture. You will have an opportunity to experience the sights and sounds of national and international significance: The White House, The Capital, The Lincoln and Vietnam Memorials, The Arlington Cemetery, and the numerous Smithsonian Museums.

We expect this year's conference will bring more international Fire Service Executives from all corners of the globe, than ever before, some of whom will share new experiences with you and discuss new ideas in Fire Protection and Safety.

IAFC '88 promises to be the best ever. I look forward to seeing you in Washington, D.C.

Sincerely, Chief Warren E. Isman, President

I was the Fire Chief of Washington, D.C., from 1982 through 1988, and had a total of 36 combined years of service. I dedicated my years of service in this magnificent city concerned with the safety and welfare of my men and women, who provide these services of safety and protection for the citizens, their property, as well as that of the city and the federal government.

At the onset of this missing photo, I grappled with the why me and what would I say to my immediate family, as to who made the decision in this matter, when was it decided, and where was the crafting of the contents done. My immediate reflection was of many years gone by, from my initial entry into the department, and the subsequent mixed encounters, my many positive achievements, gave me the solace needed to disregard this matter, although unforgettable.

When it was learned that my photo was not on the front cover of the IAFC's magazine, my top-level managers were visibly outraged, and strongly suggested the idea of stopping the convention proceedings. They felt this was done in a collective conspiracy that would ingratiate the many individuals and groups that I have had considerable disagreements with, in my management of the Fire Department here in Washington, D.C. Some of the concerns that received a considerable amount of attention was promotions of African American firefighters in particular, an affirmative action plan that was held up in administration, in a suit alleging a form of discrimination by the union local that objected to practically every decision that I made, since I failed to march to their drum beat after they had endorsed my appointment to Fire Chief. However, their endorsement had no impact on my promotion. Members of the City Council, as well as members of the Mayor's cabinet, called on the Mayor in vigorous terms to remove me from the Fire Chief position, without adequate justification, and to say nothing of the news media fanning the flame daily for my ouster, would be a gross understatement.

My Assistant Fire Chief, Ray Alfred, had received some information via telephone from a television reporter concerning the International Association of Fire Chief's Annual Conference. The information was so disturbing to the Assistant Chief, that on August 16, 1988, he shared it with a colleague in Winston Salem, N.C., Lester Ervin, Chief of the Winston Salem Fire Department.

Chief Alfred conveyed to Chief Ervin the comment made by the television reporter, "Every year since the 1970's, Fire Chief Magazine has featured in the issue immediately preceding the International Association of Fire Chief's Annual conference an interview and a photograph on the front cover of the HOST FIRE CHIEF of the city in which the annual conference is held." The question from the reporter was, "Why was not Chief Coleman extended the same treatment? Do you think it was racism?"

Chief Alfred went on to say that the reporter proceeded to further investigate this matter, by calling the IAFC Headquarters here in Washington, D.C. for answers. What he received was that the magazine was not the IAFC's. The reporter told Chief Alfred that he telephoned the editor of the magazine, in Chicago, Ill., and asked him why was there no information about Chief Coleman on the magazine. He confirmed that since the 1970's, indeed, he had featured the Host Chief in an interview and photograph on the front cover of the magazine. Additionally, he indicated to him that he was told by the IAFC that there would be no Host Chief that year. In a follow-up conversation with the reporter, he stated that nothing in the promotional material submitted to him by the IAFC, contained any information concerning Fire Chief T.R. Coleman.

Chief Alfred said to Chief Ervin, "Needless to say, it has left us shocked, puzzled, and very, very, angry. Was it a racist act on the part of the IAFC?" He went on to say that after a preliminary investigation, he had no reason to doubt the editor, and that "every preparation effort here in Washington, D.C., had led us to suspect that something was happening, but we could not figure it out, until now. The question is what will we as the Black Chief Officers, and as Black Caucus within the IAFC, do about it. This affects us all and that some answers must be demanded at the conference."

A similar letter was forwarded to Chief Edward W. Wilson of the City of Kansas City, Mo., Fire Department. His response was to Chief Ray Alfred, "I have received your letter of August 16, concerning the IAFC convention to be held in your city this month. Indeed, it has been announced that there is no host city for this particular convention. I am not aware of the reason for this, but all outward appearances would point to the fact that the Washington, D.C. Fire Department is not involved in the convention. For example, there is no opening day reception for Chief's and their wives. There is no reception for visiting wives hosted by your department's Chief's wives. There is no indication of any planned activities for convention goers after convention hours. All of these things are commonly hosted by the Fire Department of the city in which the convention is held. When I asked about these things, I was told by IAFC President, that the D.C. Fire Department was not hosting the convention this year, that it was being hosted by the IAFC staff in Washington, D.C. Perhaps you can enlighten me as to what is really going on in your city, when I arrive there later this week."

A letter was forwarded to the Chief, of Fairfax County Fire Department, Fairfax, Va. A courtesy copy was forwarded to Assistant Fire Chief Alfred, of the District of Columbia Fire Department, from William A. Anderson, Fire Chief, Lynchburg, Va. Fire Department. The letter read:

Dear Warren:

I read with much concern the letter of Chief Rayfield Alfred, D.C. Fire Department. I find it very difficult to rationalize an acceptable answer to queries of the District of Columbia CBS affiliate other than the specter of racism rearing its ugly head! If this is true, we need this as much as we need an additional hole in the head.

Just at the time the IAFC is making great professional in-roads as a voice and a force reflective of this nation's fire services, it becomes all the more unacceptable that such a slight would be made, as alleged in this letter. What an abortive event!

Having been a member of the association since 1975, and having the accomplishments of some of our leadership including our friend, the late Jim Shern, Chief of Pasadena, California's Fire Department, and remembering that Jim's photo graced the cover of one of the trade journals, the question perplexingly remains, why? In all my years there was no criteria that I can recall for this bit of recognition, other than the fact of being the chief of the department of the city in which the conference is held. Perhaps, since the locality is the District of Columbia, rather than a city, this became the failing criteria. I pray that I speak this in jest.

Ted deserves better treatment than this. I am aware, as surely as you, that Ted wanted this conference to be an unforgettable experience...but not in this fashion.

When I started in the association in1975, I can recall only two other officers that were black, as chief of a department, other than Jim Shern and Monroe Smith, of Compton, Calif., and myself. The ranks have swelled since those days. But how can this be explained to the younger minority (black) officers? If this is true, it is indeed an inopportune travesty.

You and I have been friends for years and will continue to be so; I know you, I know your family, I know your heart, I know your record, but somewhere.....in a corner, in a crack or crevice of the IAFC, lurks that old nemesis of brotherhood.....RACISM....if this be the truth. And Warren, knowing my continuing budgetary woes, I'd elected to send 4 young staff persons to the Instructor's Conference foregoing my attendance of this year's International Conference. Were it not for this fact, I'd be there to lend my voice in protest.

The Metropolitan Council of Government's (COG) located here in Washington, D.C., structured a letter of welcome from the Washington Metropolitan Regional Fire Chief's. It read, "Welcome to the region, enjoy the conference, and please take advantage of our warm hospitality and many historical sites," said Chief Harold E. (Gene) Daily, chairman of the (COG), Fire Chief's Committee and Chief of the Fairfax City, Virginia Fire Department. "This region is unique in that it encompasses the District of Columbia, Suburban Maryland, and Northern Virginia, in a geographically small, but densely populated area. We are proud to have two of our members as office holders in the International Association of Fire Chief's, Chief Warren Isman of Fairfax County, Virginia, as President, and Chief M.H. (Jim) Estepp, Prince George's County, Md., as Second Vice President. In addition, another of our members, Chief Theodore R. (T.R.) Coleman of the District of Columbia is the HOST CHIEF FOR THE CONFERENCE."

The COG Fire Chief's Committee works to develop regional plans related to fire service issues and to increase public awareness of fire safety. It also provides an effective forum for the exchange of fire services techniques and training. In view of the involvement of this organization in matters concerning various aspects of the fire department, the official officers are keenly aware of the protocol for the IAFC Fire Chief Magazine, reflecting the host Fire Chief's photo being displayed on the front cover of the magazine for the past 15 consecutive years.

The magazine made no mention of any members of the Washington, D.C., Fire Department being involved in the convention participation. On September 2, 1988, the International Association of Black Professional Firefighters, forwarded a letter to Chief Warren Isman, President, International Association of Fire Chiefs.

The members of the association committee were as follows: Chief John B. Stewart, Jr., Chairman, Hartford, Conn.; Chief Claude Jenkins, Vice Chairman, Bryan, Tex.; Chief Robert Osby, San Jose, Calif., Secretary Assistant Chief; John H. Wells, Sr., Valleys, Calif., Corresponding Secretary; Assistant Chief Richard A. Epps, Hartford, Conn., Treasurer; and Battalion Chief, Don Barlow, Anchorage, Alaska, Chaplin.

The context of the letter outlined some of the facts surrounding the omission of my photograph and comments in the magazine issue of the annual conference. It went on to state that a considerable amount of concern had been stimulated among fire department heads and black chief officers. Also, the concern was a major agenda item here in Washington, D.C., at the Black Fire Chief's Forum and at the Chief Officers Resource Committee Meeting during the International Association of Black Professional Firefighters Biennial Convention Meeting in San Francisco, Calif.

After much debate and discussion on this subject, a vote was cast to submit recommendations:

1. Withdraw membership from the International Association of Fire Chiefs.

2. Withdraw all subscriptions to Fire Chief Magazine.

3. Request Chief Ron Coleman resign his position contributing editor to the Fire Chief Magazine.

4. Request that Chief Warren E. Isman resign his position as President of International Association of Fire Chiefs.

The letter went on to state the explanation of the alleged misunderstanding between Chief Ron Coleman and Fire Chief Magazine is very difficult to believe, and that there be a resolution issued that would be acceptable to all concerned. In order to diffuse the present IAFC feelings, a strong majority mandate was structured, that the next newsletter (on scene) state a fitting apology be printed to Chief T.R Coleman, and the Washington, D. C., Fire Department, for the injustice that was created. Also, it was felt that T.R. Coleman's photo should appear on the cover of the next edition of Fire Chief Magazine, covering past IAFC convention activities.

The resolutions were viewed as not being in the best interest of black firefighters, who had struggled to get blacks to become part of the international. However, if the latter resolutions were acted upon favorably regarding the relationship with the IAFC, it was believed that the working relationship which existed prior to this incident would be restored.

In closing this communication, the wishes were to bring to the attention of the IAFC that the 1989 convention would be hosted by a Black Chief, and that prayers were such that the same misunderstanding would not reoccur. The communication was signed by the chairman of the Chief Officers Resource Committee (John B. Stewart).

On September 16, 1988, a letter was forwarded to the Chief Executive Officer, International Association of Fire Chiefs, in Washington, D.C. It read:

Dear Gary, would you please put the following resolution in the next editions of "On Scene" or in "Connection." During the 1988 Chief Officers Resource Committee Meeting at the tenth Biennial International Association of Black Professional Firefighters Convention, the following resolution was enacted regarding the growing concern for the concept of all black firestations.

1. Be it resolved that black chiefs of departments and black chief officers do not support the concept of all black firestations.

2. Be it further resolved that all black firestations constitute a return to the practice which we fought so hard to eliminate.

3. Be it further resolved that the achievement of integrated fire service is in the best interest for the professional growth and development of all black firefighters .

4. Be it so resolved this 2nd day of September, 1988.

Signed: John B. Stewart, Jr. Chairman: Chief Officers
Resource Committee

On September 22, 1988, a letter was addressed to me, at 1923 Vermont Avenue, N.W., Washington, D.C. The letter was signed by Warren E. Isman, Chief, President of International Association of Fire Chief's Inc., Address, 1329 18th Street, N.W., Washington, D.C. It reads as follows:

Dear Chief Coleman:

It has come to our attention that some chiefs have received a copy of a letter sent from a source other than the IAFC which raises some concerns about the July edition of the independently published FIRE CHIEF MAGAZINE, because Chief T.R. Coleman's photograph or comments did not appear in or on the cover as the HOST CHIEF.

Several meetings have been held between Chief T.R. Coleman, and Mr. William Randleman, editor of Fire Chiefs Magazine, and IAFC representatives to discuss this issue. According to publicity materials produced by the IAFC before and during this year's conference, Chief T.R. Coleman is the HOST CHIEF. At no time did the IAFC ever imply or state that he did not hold this distinguished title. The chief of the fire department in the city or county where the conference is held is always designated as the HOST CHIEF. If you have any questions regarding the IAFC's actions, we encourage you to read any of the publications produced by the IAFC which list Chief T.R. Coleman as the HOST CHIEF. Clearly, other fire departments in the Washington Metropolitan area have been eager to assist with this conference. In order to allow these other jurisdictions to participate, area fire departments were actively included in the planning and implementation of the conference. Chief Ron Coleman, IAFC's first vice president, in relaying this information to FIRE CHIEF MAGAZINE, during informal conversation, appears to have produced the misunderstanding that there was no HOST CHIEF. This unfortunate and obvious unintended misunderstanding seems to be what caused FIRE CHIEF MAGAZINE to use a scene of Washington, D.C., on the cover of the July issue, instead of the HOST CHIEF.

As has been planned all along, Chief T.R. Coleman has a respected, highly visible role in the ceremonial and day-to-day activities of this year's conference.

Sincerely, Chief Warren E. Isman, President---1987-1988

In the second paragraph of Chief Isman's letter addressed to me, he stated that a meeting was held between "T.R. Coleman and Mr. William Randleman." This was a complete fabrication; it never happened. One other observation: It appeared by the format and content of this letter that he did not write it. It is so unfortunate that so much time and effort were spent on the part of a few who used their positions to set the drumbeat for others to march by, and be out of sync with the desired norm.

The photos at the end of this chapter reflect the makeup of separate engine companies, by other state participants. They show the type of equipment used to extinguish fires, e.g., horses and motorized vehicles. The horse drawn method shows Engine 6, July 1864, located on Massachusetts Avenue, Washington, D.C. This was an all white male-staffed unit. Also, a similar unit photo dated June 30, 1931, shows an assortment of motorized vehicles. This was Engine 6, located on 14th and Ohio Drive, N.W., Washington, D.C.

Other photos show the fire extinguishing apparatus at Engine 4, an all black male staff unit. It was organized in 1919. This was also a horse drawn company. The motorized equipment was from years 1921-1946, located at Fourth and Half Street and Virginia Avenue, S.W., Washington, D.C.

See letter of invitation to Mayor Marion Barry, by me as Fire Chief. Also, see a similar-like cover of the magazine (IAFC) cover.

THE MISSING PHOTO
THEODORE R. COLEMAN
INTERNATIONAL ASSOCIATION OF FIRE CHIEFS
AUGUST 27-31 1998

JAN 28 1988

Honorable Marion Barry, Jr.
Mayor
District Building
1350 Pennsylvania Avenue, N.W.
Washington, D.C. 20004

Dear Mayor Barry:

The District of Columbia and the D.C. Fire Department will host the 115th
annual conference and exposition of the International Association of Fire
Chiefs (IAFC) during the week of August 27-31, 1988. The theme for this
year's conference is "IAFC 88 Future is Today". This will be the first time
in the history of IAFC that a Black Mayor and Fire Chief have served in the
host capacity.

We anticipate more than 12,000 fire chiefs, manufacturers and others who have
an interest in the fire service from the United States and aboard to attend
this event.

The President of the United States, representatives (who will probably be
ambassadors) from over 40 countries along with fire chiefs from their respective
countries, have been invited to attend the conference.

As host Fire Chief, I have the pleasure of inviting you as Mayor of our great
city to participate in this historical occasion in the Nation's Capital. Your
participation should include greetings and mayoral remarks at the opening
session, tentatively set for late afternoon, Sunday, August 28, 1988, at the
Grand Hyatt Hotel, 10th and H Streets, N.W., in the Grand Ballroom.

On Wednesday, August 31st, you are again invited to join us for our annual
banquet and President's reception. The President's reception will begin at
6:00 p.m. and the banquet will start at 7:00 p.m. at the Grand Hyatt Hotel, in
the Grand Ballroom.

Please find enclosed for your information, a copy of the conference agenda. If
there are further questions relative to the invitation and request, please do
not hesitate to call. My office and staff is at your disposal.

Your immediate response will be greatly appreciated.

Sincerely,

T.R. Coleman
Fire Chief

Enclosure

GS:dlc:1-26-88

cc: Files/PA
     FC
     Alfred
     McCaffrey

170

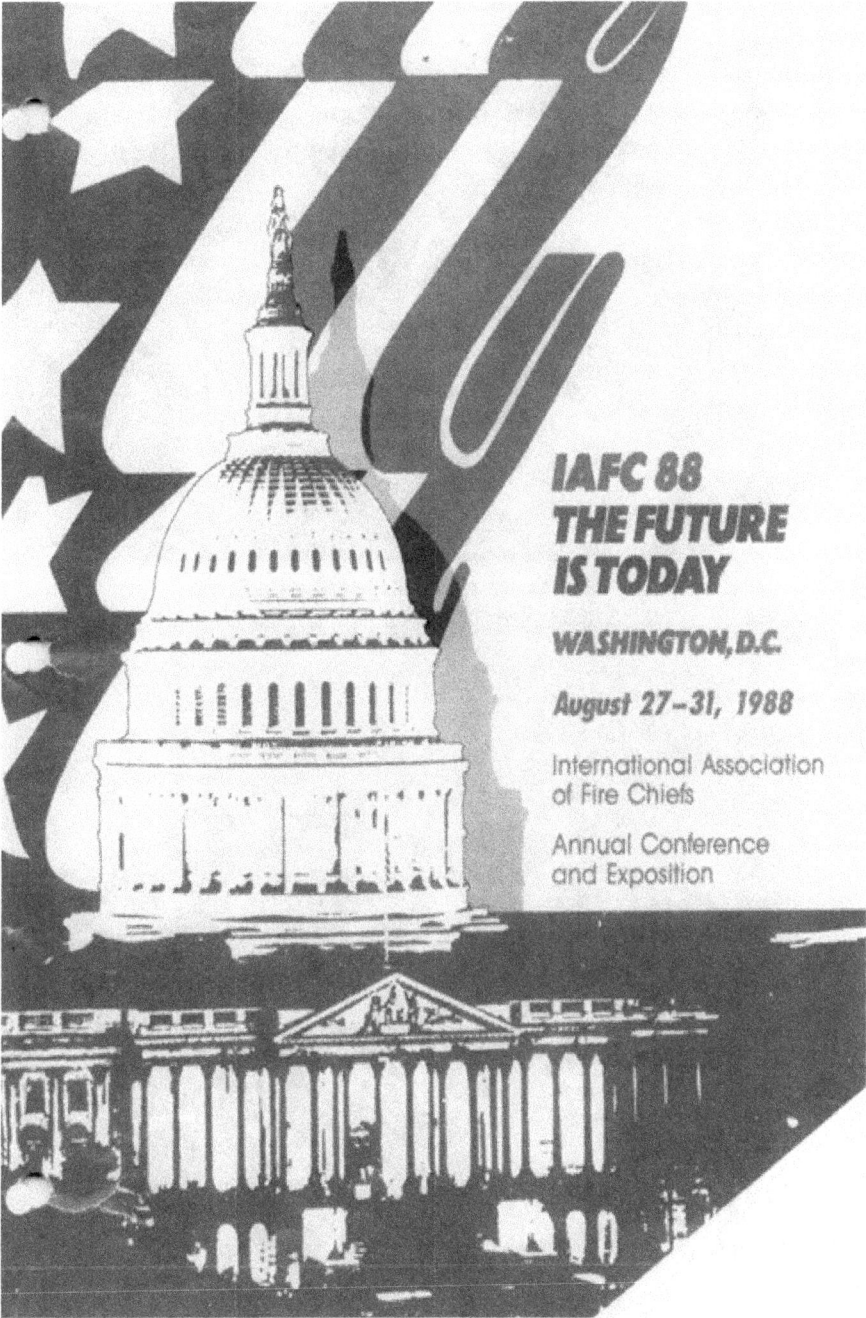

IAFC 88
THE FUTURE
IS TODAY

WASHINGTON, D.C.

August 27–31, 1988

International Association
of Fire Chiefs

Annual Conference
and Exposition

Chief T..R. Coleman and Family Enjoying The Parade

Participants In The Fire Fighters Parade

177

The Ronald McDonald Train in the Parade

Junior Fire Fighters

Apparatus Sporting a Dog (Dalmatian) in the Parade

Newburg Hopewell Volunteer Fire Company Participating in the Parade

THE ROUGH RIDER

A MILITARY TYPE ARMORED VEHICLE MODIFIED WITH A 1/2-INCH NOZZLE, 2 1/2- INCH HOSE, WITH A 90-LB. PRESSURE STREAM. IT WAS MODIFIED BY BATTALION CHIEF TATE IN OUR APPARATUS DEPARTMENT.

FUN AFTER THE PARADE
CHIEF COLEMAN ISSUING AWARDS TO WINNERS IN THE APPARATUS CONTEST

July 1, 1864
Engine Company No. 6
438 Massachusetts Avenue, N.W.

ENGINE Co. 4
1919

DCFD Engine Co. 4 Personnel and apparatus when organized in 1919 as an all African-American company. FFA collection.

Engine Co. 4 circa 1921 with first motorized apparatus. On left: 1915 Christie Front Wheel Drive Tractor attached to 1907 American LaFrance 700 gpm Steam Fire Engine. On right: 1921 Brockway 50 gallon single tank combination chemical hose wagon.

June 30, 1931
Engine Company No. 16 - 14th and Ohio Drive, N.W.

Chapter 25

Retirement Decision and Retirement Ceremonies

In the latter months of 1988, I pondered the idea of retiring from the District of Columbia Fire Department by year's end. The reasons were only a few: My wife Uvaghn had supported me during my ups and downs in my issues with the department in the various segments of leadership, and I owed her my appreciation. Also, I felt that there were persons in leadership positions that were capable and qualified to maintain the level of excellence that had become the modus operandi that they had helped me to establish; and lastly but not least, when accumulating my physical years, it was time. Having said that, I contacted the Mayor and informed him that I was submitting my papers for immediate retirement, for his approval. His response was slow in coming. He finally asked me, "Are you sure you want to do this"? I answered, "Yes, Mr. Mayor." He responded saying, "Consider it official. I will sign your request." I called a meeting with my staff, and made the announcement.

Celebration and Farewell Speeches

Color Guard
The National Anthem

The Opening Ceremonies by Joseph Yeldell

Good evening, we are here to celebrate and indeed to celebrate a young man retiring in his youth after 36 years of dedicated service to this city, Washington, D.C.

I know you are going to welcome him a little later, let's give him a round of applause.

I want everybody tonight to undertake where we are and what we are doing to let him know how much we love him, how much we respect him, how much we appreciate what he has done, and that loyalty stays with him as he moves into a well deserved retirement.

I was going to introduce the head table at this time, but we are waiting on a few more guest, so I am not going to do that. So what I'm going to do then, is take the time, just before we have dinner, and ask the Chief's very, very, lovely family to stand.

We are going to move ahead with the planned program. At this time we will hear from our City Mayor, Marion Barry.

Marion Barry

Thank you very much Joe Yeldell. Chief Coleman, Mrs. Coleman, and entire Coleman family, the D.C. Government is very delighted to be here with you tonight.

Chief Coleman, it is a pleasure to join all of you tonight to celebrate your 36 years of dedicated service. Thirty-six years of dedicated service is a long time to do anything. The same job. I don't intend to do this job this long.

I believe that public service is a privilege, not a right, and that we all should look forward to doing our mission, and being a Fire Chief in some instance of mission impossible, being a fire Chief of a large major city.

I believe that part of our public service is respect. We may not like that person as much as we would like to, or agree with him or her as much as we ought to, but the institution is worth respect. It seems to me that every Assistant, every Deputy, and every Battalion Fire Chief should be here tonight to honor our Fire Chief. Our Fire Department is one of the best in the nation. I would like all fire department personnel, as well as civilians, please stand up. Thank you, and Chief Coleman thanks you too.

Chief Coleman got this department at a time when segregation and discrimination was rampant. Black firefighters slept on one bunk that had their name on it, and they couldn't move from it. The plates they ate out of would be broken up. There were times when you were called names that were not complimentary. Yet in spite of all that kind of treatment from many of his fellow white firefighters, and officials, he endured; he endured. A lot of people with less character, less integrity, would have been so bitter, and so wrapped up in that kind of situation, they wouldn't see anything else but that. The character of this man, the integrity of this man, helped to overcome that, where he sees all members of the fire department, Black, White, Seminoles, Jews, Protestants, Catholic, as part of the government, and it's importance, the way it should be in serving our community. They are also the ones that made the final difference. He made sure people running for positions have an opportunity to perform. Let us give him a round of applause.

I remember in 1982, I had the difficult task of selecting our Fire Chief, and you all know my footpath, I take them all the time, with broad shoulders, and belief in God. In 1983, I made him permanent. In the very beginning, the appointment was controversial.

You all know in1987-1988 my election was coming up. Look at downtown now; look at our neighborhoods now. Our unemployment is the lowest in 25 years.

I made a point that when travelling unchartered waters, you are bound to hit a rock, go aground. Whoever heard of a boat moving forward without making waves. Chief Coleman has made his share of waves, but he also moved forward too. This department has outstanding firefighters, medical personnel, and others. Chief Coleman has moved us forward in initiating the smoke detector program, the first of this kind, I think in the nation.

Firefighters when they are not fighting fires, they are out inspecting buildings, they are out inspecting homes, and apartments, teaching people fire safety. The lowest fire deaths in I don't know when, under his leadership. I think last year, about 10.

Not only was he advocating programs, I recall when they started the smoke detector program. Chief Coleman and his wife were right out there talking about smoke detectors, and giving to those who needed them, as well as helping people to install them. That's the kind of leader we need, that's the kind of public servant I believe in.

Chief Coleman needs to be judged by the whole person, not like a prizefighter, or a football game, or baseball game, based on the number of innings or quarters. Chief Coleman had to look at the whole season. One day you might hit a home run, a triple, you may strike out, it's the whole season that counts. If each of us would be judged by every individual act, rather than the whole season, we wouldn't be where we are. I judge this man by the whole season. Thirty-six years of service, 6 years of being my Fire Chief. Let me tell you, he hit more home runs than he struck out. The city hosted the International Firefighters Convention. Thirteen thousand came to our city, the last conference we ever had in this city by the best Fire Department, and the best Chief. You should have seen the grin on the firefighters' faces that did that.

Chief, I know we have been through thick and thin together. I know that you are happily retiring. I am going to have some assignments for you, outside of your playing golf, or fishing, anybody can do that, that know how to fish. I am going to call you from time to time for volunteer work (pro bono). I am sure I am going to hear from you. I hate to see you retire. I know that 36 years is a long time. This is my 28th year of public service for the public, you do want to do other things sometime, but I like what I am doing, and I hope you all do, too.

Chief, I am going to ask you to come up for a minute. I would like to present to you this distinguished Public Service Award. I would like my cabinet members to stand. I want you to see them too; 45 of them are here tonight.

Theodore Coleman, in recognition of your outstanding contributions to the government, and the people of Washington, D.C., for a period of 36 years, an important era in the history of this great city.

There are hard jobs in the D.C. Government, they are all hard, some are harder than others. Chief Coleman, it's been a great pleasure being close working with you. One thing about the chief, when he was in office, he has been loyal so far. Some of you may have been upset with the Chief about a few things, but he was carrying out my orders. I was involved with that, not the Chief, this was for business reasons, things that I couldn't say. Anyway, Chief we love you. God bless you, you have been a faithful servant, we are going to miss you. I am not going to talk too long, this is a celebration. And finally, it's good to see you occasionally out of that eight-button box—ha ha ha. Any time you saw Chief Turner of the Police Department and Chief Coleman in public, they were wearing those eight-button boxes; it was a beautiful sight.

Chief, I am going to ask you to come up for a moment again. I would like to present to you this outstanding Public Achievement Award for your outstanding contribution to the government and the people of Washington, D.C., for a period of 36 years, an important era in the history of this great city. You provided positive leadership for the Fire Department and City Government. You progressively pursued solutions of a wide area of complex problems affecting the quality of life of citizens in Washington, D.C. You have confronted those challenges with pronounced profound professionalism, with personal commitment and deep sensitivity, for the concerns of these most affected. You have earned the respect and appreciation of your colleagues and this certificate for qualifying your services.

Marion Barry
Mayor

Our next speaker is the former Fire Chief, 1973-1978, Burton W. Johnson. He was the first black chief in the District of Columbia Fire Department.

Burton W. Johnson

Distinguished Master of Ceremonies, Chief Coleman, Mrs. Coleman, platform dignitaries, ladies and gentlemen, good evening. The Master of Ceremonies has introduced me as the representative for the fellowship of Retired Police and Fireman. Additionally, after the presentation for the fellowship, I'll have to put on another hat.

Chief Coleman, I have a certificate from the Fellowship of Police and Fireman Association, reading thusly:

Theodore Coleman, Chief of the D.C. Fire Department, on this date, November 29, 1988, we take great pride in saluting you. We have known you and served with you for many years. We more than know what you have endured as a firefighter over the years. It is common knowledge among us that the stressful conditions under which you worked within as well as on the fireground. We also realize that traditions are hard to change, whether they are negative or positive. We have experienced many changes in attitudes of our society. Through the good graces of God, you've endured all these challenges, and successfully held the highest post that is available to a fireman and, you did a great job. We just want to let you know that we hold upon you in high esteem, and wish you and your family a very wonderful retirement, full of peace. On behalf of all officers and members.

Chief Coleman, President Paul V. Brown has also asked that I make a presentation to you of a ticket which entitles you to be admitted to a luncheon on December 2, 1988, at 11:00 a.m. The luncheon so indicated that fellowship is casual, and to also read to you another letter indicating on behalf of all officers and members, we would like to extend to you one full year's membership in the fellowship of Retired Police and Fireman of D.C. He's asking that you fill out this application for membership. But in anticipation of your presence and your joining the organization, we have also brought with us a badge and pocket emblem which now indicates that you are a member of the fellowship of Retired Police and Fireman.

Before I move on to the next issue, let me ask all of the Retired Police and Fireman's Organization, please stand.

Now comes the other part. Chief, I am going to ask you to stand again. I have a badge that several of your friends, companions, buddies, and whatever else you might want to call them have empowered me to present to you, and it is a token in recognition of your significant contribution and outstanding dedication to the District of Columbia Government and its citizens.

Chief Coleman, a pocket badge indicating your appointment date January 5, 1953, your promotion date, April 19, 1982, and your retirement date, December 1, 1988. Chief, congratulations, and will all the men who were serving at Engine 7, Truck 10, on the date when Chief Coleman came in as member of Engine 7, please stand. You served in this department for 36 years.

Our next speaker is Chief Jim Estepp of the Prince George's County Maryland Fire Department.

<u>Chief Jim Estepp</u>

Ladies and gentlemen, Chief Coleman, it is my pleasure to be here. I've got to do something before I make a remark about you. I would be remiss if I didn't make a remark about someone else here in this room.

On my way over, I heard a very contemporary song by an artist by the name of Jerry Morris. He's a good country, western singer, if you like country, western music. The words kind of go like this. It makes me think of somebody in your life. If he were standing here right now, he would probably say the same thing. It goes like this: *"You know that you're my hero, you're everything I want it to be, I can fly higher than an eagle, because you are the wind beneath my wings."* Uvaghn, I think everybody feels that way about you, the support you gave Ted. I know that he would say the same thing. I want to say that first to you.

On behalf of your colleagues in Prince George's County, who worked with you for a number of years, and on behalf of your colleagues in the International Association of Fire Chiefs, it is my distinct honor to be here. There are 33,000 fire departments in this country and 1.5 million firefighters. Firefighting is one of the most dangerous occupations in this country, and this is evident by the six firefighters killed this morning in Kansas City, right up there certainly with the brothers and sisters in law enforcement.

There are a number of firefighters who are killed. There are a number of Fire Chiefs that are divorced, in fact the statics are that Fire Chiefs have just about the highest divorce rate of any public service in this country, and you're fortunate Ted that this has not happened to

195

you yet. So you are retiring at the right time. But for all of us in this room tonight, and others who have served in Chief positions, you all know what I am talking about. It's an extremely difficult job; you win some, you lose some. But always there is the mission that comes first, and that's to save lives, and property; and it appears that you have done that, because of that work you have done, because of the dedication and pioneering, you carried out an extremely difficult job, often under extremely difficult conditions.

Your colleagues in Prince George's County would like to make this presentation to you.

Chief T.R. Coleman, in recognition of your many years of dedicated service to the citizens of the District of Columbia, for your outstanding cooperation with the Prince George's County Fire Department.

Please accept this as a token of our appreciation.

One more, the Mayor mentioned that last year, the International Conference of the Association of Fire Chiefs conference was held here in Washington, D.C. The Mayor never ceases to amaze me with his ability for covering statistics, and he was right on the mark. This was the largest attended conference that we have ever had in the history of the International Fire Chiefs Association, right around 13,000; and without a doubt, it was the most successful conference in our history, and we have already as a result of this conference sold exhibit space in Indianapolis next year in excess of what we did here, because of that success. And on behalf of over 10,000 members of the International Association of Fire Chiefs, because of the outstanding job you did here in August; on behalf of our officers, our board of directors, and particularly Rick Holdman, our president who is in California, and could not be here tonight, Ted please accept this distinguished service award from the IAFC and all your colleagues.

Our next speaker will be the Washington, D.C., City Administrator, Mrs. Carolyn Thompson.

Mrs. Carolyn Thompson

Thank you very much Joe. Good evening to everyone. To Fire Chief Coleman and table guest, ladies, and gentlemen. It is my honor to join you this evening to celebrate Theodore R. Coleman, our Fire Chief, in 36 years of service to the District of Columbia.

Most of us know Chief Coleman. He contributed exemplary performance as a firefighter for his trusted friends, but all of us know that his commitment to the District of Columbia is unsurpassed.

I worked with Chief Coleman and some department heads some years before I became City Administrator. At the time, we worked together on some very difficult problems such as especially eliminating the certificate of occupancy backlog and the need for updated codes in the District of Columbia. Since my appointment in February 1988, I have grown to understand more clearly the tremendous challenge involved in the day-to-day operation of the Fire Department, and I have gained tremendous respect for Chief Coleman. He fought daily, and I have to reiterate, fought daily against all types of obstacles.

Some of us have had the opportunity to know Chief Coleman in the workplace; some involved in a special organization, such as the International Association of Fire Chiefs, International Association of Black Professional Firefighters, the Council Government, and the National Forum of Black Public Administrators. Others have known him through his civic work, such as the Metropolitan Police Boys and Girls Club; and we all know that he has given a great deal to our community, and we are all richer because of his involvement.

During his tenure, our Fire Department was revitalized and modernized. Because of Chief Coleman's input on fire prevention in 1987, our city had only 10 fire fatalities, the lowest number in 50 years. Chief Coleman takes great pride in his remarkable track record. As he leaves us, he challenges all of us to help maintain and improve upon his benchmark. I know all of us, employees, citizens, and faculty members alike will miss Chief Coleman.

He is a valued and appreciated member of our family. I commend you for a job well done, in leading the D.C. Fire Department, which is a first rate team, and earned the respect of the residents of our city.

Chief Coleman, thank you for all that you have done, and best wishes for a long happy retirement, and please enjoy your evening.

Our next speaker is Mrs. Alexis Roberson, Director of Employment Services.

Mrs. Alexis Roberson

Good evening, Chief Coleman and Mrs. Coleman, the Coleman Family, ladies and gentlemen, good evening. I would like very much for the Chief to stand. It is indeed a privilege for me to present this tangible token of our appreciation for your outstanding stewardship, and as an outstanding Public Servant. It's a privilege for me to tell you that, because you are a leader and a good friend; but more importantly, because you are a real home boy. We're both from a little town the size of this ballroom in S.C. I didn't know all of his family members, but did know his father. So it is a privilege for me to give you this little token, and I'll read it quickly. It says, *"Best wishes to a friend, and cherished colleagues, Chief Theodore R. Coleman. As you retire from the District of Columbia Government, I would like to take this opportunity to salute you and to wish you well in the years ahead."*
Chief Coleman, you will always be missed, and more importantly, you will be remembered for your style, for your warmth and human understanding, for your devotion to the District of Columbia Government, for the citizens it serves, and years of distinguished services. History will record you as one of the city's finest Fire Chiefs. It will also show the progress the department made in upgrading the overall Fire Safety Program under your outstanding leadership. Indeed, your career epitomizes how one can rise from the bottom rung of the ladder to the position of prominence.
Chief Coleman, although you are leaving government service, I will regard you as a trusted friend, and a cherished colleague. I wish you many years of continued good health and happiness, and I congratulate you and your family on your distinguished public service career. Survival business will remain immeasurable, and will serve our city for many years to come. Indeed, on this occasion of your

retirement, the Department of Employment Services is pleased to salute you for your many years of loyal service to the citizens of our great city.

The next speaker is Battalion Chief Cornelious Jackson, representing Greater First Baptist Church, on Thirteenth & Fairmont St., N.W., Washington, D.C.

Battalion Chief Cornelious Jackson

To the Honorable Marion Barry, Mayor of the District of Columbia, to the retiring Chief, T.R. Coleman, members at the head table, and friends who are sitting out on the floor.

First of all, Chief Coleman, I bring you greetings from my church, Greater First Baptist Church, on Thirteenth and Fairmont. They asked me to wish you a happy retirement, and may God bless you. My pastor is Dr. Edward Thomas. He has followed your career, and has kept up with what's going on in the newspaper. He said you are a strong man. The people in my district, they appreciate things that you have done, while you were Chief of the District of Columbia Fire Department.

I met Chief Coleman 26 years ago. When I met him, I was coming on as a rookie firefighter. Chief Coleman, Joe Kitt, Maurice Kilby, many older firefighters, we looked up to them, I looked up to them, and I was interested in being an officer. They told me what I had to do to become an officer. I remember as if it were yesterday; Chief Coleman gave me a first aid book to study. He said, you got to know this, and I appreciated this.

Chief Coleman, for the past thirty some years, you have been a faithful servant of the District of Columbia Government, and as Chief of the Fire Department for the past few years. Several weeks ago, I looked at a certain article in print that said, "The controversial Chief of the District of Columbia will retire on December the first," and what it should have said, "The Fire Chief of the District of Columbia is retiring."

Chief Coleman, during your tenure as Fire Chief, you've done an outstanding job. Through the many complex problems of the Fire Department, you still moved the Fire Department in to the twenty first century. You have gotten modern equipment, instituted the Home Safety Program, and you've seen to it that all Battalion Chiefs in their districts attend the community meetings, so they could answer any questions the citizens of the community would ask. This is the first time any Fire Chief has ever done that.

The citizens, especially on Capitol Hill, in my battalion, realize that. Also, they had many questions of the Fire Chief, of what was going on in the department. But one thing Chief Coleman, of all the good things you have done, very few of the good things were published by the media. It seems as though the media has invented the game called let's bash the Fire Chief of the District of Columbia Fire Department. We'll do this by not reporting the good things he has done, but we'll report everything negative that we can think of. And it was hard to be watching TV at night, it was always a source that didn't want to be named.

You know, several days ago I had a pick-up-truck pass me that had a sign on it that said, "Whiskey bent and hell bound." Every time I heard someone on TV saying that something was wrong with the District of Columbia Fire Department, but they didn't want their names revealed, that's what it reminded me of, the guy with the sign on the back of his pick-up-truck. "Whiskey bent and hell bound" is determined on destruction. And these people who contact certain media, and won't give their names, they are bent on destroying the District of Columbia Fire Department.

So, Chief Coleman, now that you are going to retire, I want you to know that I've often said this when I saw you, that you were an outstanding Fire Chief, you were a good Fire Chief, you saw to it that the men get adequate training, and the guys in my battalion always talk about Chief Coleman, and what they thought of Chief Coleman. I want to say this Chief Coleman, congratulations on your retirement, and may God bless you.

The next speaker is a representative with a message from Ms. Carol Hill, the Executive Director of the D.C. Commission for Women.

<u>Ms. Carol Hill</u>

Good afternoon. At this time, I would like to present to the chief, a message from Carol Hill, the Executive Director of the D.C. Commission for Women.

Dear Chief Coleman, please allow me to convey best wishes an appreciation to you on the occasion of your retirement. Your years of dedication in public service is invaluable to the citizens of the District of Columbia. Because one of our primary concerns has always been employment opportunity. I would be remiss as the Executive Director of the D.C. Office of Commission for Women, if I didn't offer a thank you for not only the number of female employees hired in the Fire Department under your leadership, but for the number of women who have been promoted under your direction, in recent years.

The gratitude shared by all those under the managers committee which comes under the Office of Commission for Women, I am sure the women of the Fire Department, who have served under your tenure as Chief, extends best wishes on a well deserved retirement, for your future, and a world of success.

The next speaker is Assistant Fire Chief Ray Alfred.

<u>Assistant Fire Chief Ray Alfred</u>

Good Evening, and I promise I am going to be brief.

Chief Coleman, on behalf of the department, we have a plaque here, and, I'll simply display it. We'd like to show our appreciation for your leadership, and by another plaque that says, *"Thanks for taking us from being ducks to become eagles."*

And then finally, traditionally, when one of our members retire, especially for someone who has gone through the ranks, we traditionally give them all of their badges, from the day they were appointed, to the last rank they have held; and we do not want to break a tradition. Chief Coleman started with badge 584, which was the badge assigned to you as a private, not a firefighter. On behalf of this department, we congratulate you and wish you well, and we appreciate the years of your leadership.

The next speaker, Mr. John Lund, representing the Friendship Fire Association.

Mr. John Lund

Chief Coleman, I just wanted to extend greetings and many thanks from members of the Friendship Fire Association. For those who don't know what we do, we maintain canteen services for Truck Engine 31 on Connecticut Avenue. The most important service is providing volunteers to maintain canteen units, which the Chief has been a tremendous supporter. Mind you, it is not as important as a ambulance, ladder truck, pumper wagon, or whatever you want to call it, it is an important service, and those of us that have wives or girlfriends, find it hard to understand a bunch of people getting up two or three o'clock in the morning to a multiple alarm blaze, and somehow manage to get to the office the next day. We try to do that, and the Chief has attended several of our meetings, he is very supportive. We want to thank you for your efforts in finally getting us a canteen unit, and we hope to get it in service in not too distant future.

Chief, I don't know where you're going to hang all these plaques. I would like to present to you this plaques from the Friendship Fire Department: The Friendship Fire Department bestows upon the Honorable Theodore Coleman, Fire Chief of the District of Columbia Fire Department, an Honorary Membership, effective November 21, 1988, for his continuing support. Healthy and Safe Retirement.

The next speaker will be, Mr. Eugene Kenlow, representative for the Eastern Branch of the Kiwanis Club.

Mr. Eugene Kenlow

Master of ceremony, to Chief Coleman and Mrs. Coleman, Mr. Mayor, and distinguished members of the head table, ladies and gentlemen, good evening. I didn't ask for this, I was told to do this. So a command performance is expected of me. There are some people across the river who said, "We know a man who is a good man, and since you are the president of the Kiwanis club, you have to go and say thank you.

So it is my great pleasure to come this evening to say to Chief Coleman that we remember that every year for a lot of years, particularly the last 6 years, there is an annual Christmas party out at Burrville Elementary School for seniors, and tots, and for people who would not have Christmas that would be memorable, if it were not for the Fire Department under your leadership. And that said to me that the test of the individual is shown by his or her care for those who are in the dawn of life, and you demonstrated more than just about fire safety, emergency medical services. You demonstrated that you care about the fullness of life of all people. And we can always remember that ever year around April or May, we had our annual awards program to try and honor our outstanding citizens, people who have been heroes, role models, in our community, teachers, police, firemen, that you are always there, you are always supportive.

No, I didn't bring a plaque. See I am from across the river, but I did bring people, at least they brought me. And lots of our people wanted to come to pay tribute to you, but see I didn't bring a plaque, what I did bring was an invitation. What we are going to do at a time when it is mutually convenient is to get you to come back home across the river on our turf, so we can say thank you appropriately. Is that alright? Okay.

They tell me the genius of effective leadership is somebody who can put the right person at the right place at the right time. My friends tell me when history is written, that in retrospect, someone is going to call you a genius for making this appointment, Mr. Mayor.

Finally, T.R., they tell me in 36 years, you put out a lot fires. Based on the news in the past 24 hours, I think you are going to put out one more. Of all the fires he put out, as they say, "No problem, don't worry, be happy."

The next speaker is the president of the International Association of Black Firefighters, Lieutenant Romeo Spaulding.

Lieutenant Romeo Spaulding

Good evening and almost good morning to the Honoree tonight, then distinguished Fire Chief of the District of Columbia Fire Department, the Honorable T.R. Coleman, his wife Uvaghn, to the master of ceremony, the City Administrator, the Honorable Carol Thompson, and other platform guest, friends and relatives of the honoree tonight. First presentation that I am going to give to you is one given from the Community Relations Unit of the District of Columbia Fire Department. It's the one that those in the fire department don't know about, and the media just found this out a couple days ago, and put us on the front page of the Metro. But the fact is this unit has been going out carrying out the mandate of the one that we are honoring tonight, going out into the community trying to educate our community on fire safety methodology. And of all the programs that you have heard delineated tonight, it was a direct derivative of the man we are honoring tonight, T.R. Coleman. He is a person that chartered those causes, and he is also the individual that gave the green light to go out in the communities and do the kind of work that was done. And of course, as head of the fire department, he is the person that received the credit for doing all that, for without a man with a vision, without a man that doesn't have a commitment to serve his people, that people will not receive quality service. So the Community Relations Unit, Chief Coleman, would like to honor you by just saying one thing and I'll give it to you tonight. I was the only person that was threatened before I got up here. I am going to have my legs broken, and I am going to be choked and everything for not being short. I am also the only person that has been identified by the media as Chief Coleman. They have printed my picture in the paper and his picture with our names you know, transposed and as I took my hat off, he put his hat on; I had to change glasses because when I went into Ward 3, they did not know that I was not Chief Coleman. But that's okay; Chief will keep the records. The Community Relations Unit would bring Chief Coleman on and he will actually tell you about how he would go out into the community, because this man has been the Fire Chief of this Fire Department of the District of Columbia, but it is part of our nation's capital. When you start equating the fire department with another department across the nation, this includes Prince George's County, Fairfax County, and all surrounding counties, and every city. It is the pentacle of fire departments in the nation and also in the world.

Every Fire Chief who has been a chief in the International Association of Fire Chiefs would love to have on their resume' that they were the Chief of the District of Columbia Fire Department. So what that says to Chief Coleman is that he was not only our fire chief and he was not only the number one fire chief in the nation, but he was number one fire chief in the world. This would also have shown the Mayor of the city as being black, as well as the Fire Chief. And I will only tell you that the reason why that is not the echo more is because he happened to be born the same type person as Jessie Jackson. The plaque reads thus. It says, *"To the number one fire chief in the world, T.R. Coleman, thanks for your excellent leadership. May your retirement days be filled with love, joy, peace, health, and happiness, the staff of Community Relations Unit of the D.C. Fire Department, presented November 29, 1988."*

As I switch up, and transfer my hat, I do have another dubious responsibility, and that's being the president of the International Firefighters, which numbers in the organization of 8,000, approaching 9,000, and if you take in the fact a 170 some thousand firefighters in this country became professional firefighters. We make up approximately 11.6 percent of the total, and to just correct you on a couple of things, the District of Columbia fire department does not have the largest number of female firefighters, we are number one in black female firefighters in the nation. (Applause). It gives me an honor to be standing tonight on behalf of those members of the International Association of Black Professional Firefighters. And to stand in honor to recognize a member who has kept the fire burning for justice, which is our motto. He has stood bravely the test here in our nation's capital, and he has fought courageously the battles that have confronted him. He is a man of unwavering character, and even as you have heard tonight, how he has maintained his standing. We are not holding on to Chief Coleman or "T.R.," as he is called, to retire and not be active, but to be a continued productive citizen in this community, as has other chiefs who have gone on before him. These are the kinds of things that we call the members of this association to. One thing I would like to say to the chief that as we look on his resume', and, listen to what they are saying about you and your affiliation, has listed the International Association of Black Professional Firefighters, that you are proud to be a member of. At this time chief, on behalf of International, I would like to present you with this plaque.

It says, *"A Letter of Retirement Expression. This letter of retirement expression is presented to Fire Chief Theodore R. Coleman, Fire Chief of the District of Columbia, who has been a member of the International Association of Black Firefighters for the past 18years, that's as long as this organization has been in existence. He has exemplified the type of commitment and dedication that is necessary for one to achieve the highest rank in the fire service, and that is Fire Chief of our nation's capital Fire Department. Chief Coleman has completed over 36 years of honorable service as a uniform member of the D.C. Fire Department. He is also a strong supporter of family and education excellency which is the most needed commodity in our day and time. He has served the cause of justice, peace, and freedom well during his tenure as a member of the association, and Chief of the D.C. Fire Department. On behalf of the IBPF membership, may you enjoy a very fruitful and joyous retirement."* I would also like to say to Mayor Barry that we would like to thank him for clearing up some things, and I think everyone here is aware of what I am talking about, in that he said the Chief was a loyal servant in following his directions on a lot of those things, and we all knew that. Right?

I would like to say to Mrs. Coleman, thank you for standing with the man, it was not easy, and the accomplishments that you made Chief, we realize that now. We have more females in the Fire Department than any other Fire Department in the nation. Chief you have promoted numerous people; what else can I say. As I told you earlier, and I hope I don't offend anyone by saying this, "You're a hell of a man."

Chief Coleman, we present you this plaque, and it reads: *"Black warrior for 36 years of service in the fight for equality."* Believe me, he is a class of honor.

The Mayor mentioned the Chief taking over the department, and that it was a difficult time, and that America was a difficult place. Some of us say that America hasn't changed that much. I would be the first to say it hasn't. Chief Coleman came along when segregation and discrimination was rampant.

Booker T. Washington once said, "Success is not measured so much about a position that one has reached in life, but by the obstacles that he had overcome in order to achieve." No doubt in my mind, Chief Coleman went around obstacles, jumped over some, and under some. As we sit here tonight trying to glorify this human being in his last hurrah, and yet it's not about him, I think it's called the Washington Post. Chief Coleman, it has been an organization privileged to bestow upon you, a retired Fire Chief's Honorary Membership in the Progressive Firefighters Association, and with this sir, the military fashion that you pin the stripes on. We will pin a lapel pen on you.

Also, this is a plaque that reads of our appreciation: *"Presented to T.R. Coleman, in recognition of outstanding service, from the Progressive Firefighters Association, of Washington, D.C., on this 29th day of November, 1988."* President—Chairman of the Board

We will listen to a statistical tribute to Chief Coleman by one of his special assistants, who believed him to be the greatest, Ms. Tisa B. Crutchfield.

Ms. Tisa B. Crutchfield

Fire Chief T.R. Coleman entered the D.C. Fire Department 36 years ago, at the age of 26. After years of progressing steadily through the ranks, he is now at the top of the ladder, the D.C. Fire Department's leading Administrative Officer. To his present position, he brought with him more than a quarter of a century of lessons learned and methodologies developed to cope with changing situations.

One of the most important things he learned was to be flexible. Times change, and people must change with them. Under his leadership, many positive changes have been made in the fire service some of which I would like to share with you, the public.

As the District's fourth black fire chief, Chief Coleman has experienced prejudice during his early years, and today, much prejudice yet exits. He is working diligently to reduce this social cancer through skill and perseverance. One of his greatest philosophies is that communication is a wonderful tool which must be used to help break down racial barriers in any organization. He welcomes the day when blacks, whites, and other ethnic groups in the Fire Department will start relaxing and listening to each other and respecting other groups' different views and life styles. There is definitely a lack of understanding among these groups. We should get to know each other and set at a stage where in our department, people accept each other not by color but by ability to perform a job. This would be gratifying to see.

Chief Coleman has not been standing still during his tenure as Fire Chief. He has drastically realigned the administration internally, and has also created numerous technical positions, namely: Resource Specialist, Judicial Affairs Officer, Medical Doctor, Labor Relations Officer, Special Assistant, Operations Research Analyst, and Compliance Officer. Chief Coleman has broken the record of his predecessors for his innovative ideas and for making things happen.

The following list speaks of only a few of his accomplishments:

- Institutional Fire Safety Program - This is a collection of seminars and workshops designed to enhance the knowledge of staff and residents of institutional facilities with regard to fire safety survival.

- Fire Safety System - This system is designed for District residents to broaden their prospective on how to stop fires before they start. It is composed of 3 components: (1) A Smoke Detector Giveaway Program, (2) A Fire Prevention Education Program, and (3) A Home Safety Check Program. These programs are offered through the Community Relations Unit of the Fire Department. This overall system is credited for the number of 10 fire related fatalities in FY-87, being the lowest number that the District has had in 50 years.

- Analytical Task Force - Designed to increase efficiency in our Communications Division through a Computer Aided Dispatch System and to identify the need of Automatic Data Processing equipment and personnel.

- D.C. Firefighters' Parade - An annual event which symbolizes heroic acts of bravery and dedicated services performed by firefighters and which also aims at making Washington a fire safe community.

- Community Action Team (CAT) - A team which coordinates activities of all Battalion Fire Chiefs for ensuring continuous contact between the Fire Department and Advisory Neighborhood Commissions, Civic Associations, and other ethnic groups to exchange ideas in ascertaining the needs of all parties involved.

- Honor Guard - Established in 1981 to accord respect to deceased members of the fire suppression force, both active and retired. The guard averages 20 ceremonial functions per year. Other functions include helping to open and close the Special Olympics, the National Olympics, marching in the D.C. Firefighters' Parade, and participating in departmental ceremonies.

- Color Scheme - The decal of red stars and bars on a white background was the color scheme originally designed to be placed on all fire apparatus in 1982. The Mayor has since adopted this color scheme to represent District vehicles in local government.

- Mass Casualty Metro Drills - A major simulated exercise was conducted on November 7, 1982, to test and evaluate the mass casualty incident plan, standard operating procedures for Metrorail incidents, pre-hospital care for sick and injured including actual transport of over 200 victims to 10 area hospitals, proposed medical equipment supply vehicle, and the mutual aid assistance of 5 Montgomery County, 5 Prince George's County, and 2 private ambulance units.

- First Command Vehicle - A converted medic unit equipped with telephones and radios for continuous communication between units on the fireground and the Communications Division.

- Armored Personnel Carrier - Nicknamed the "Rough Rider," this unit is equipped with fire suppression capabilities for attacking hostile fires when working under gun fire conditions. This is the only APC in the country modified for firefighting use.

- Military Army Insignia - Designed to give immediate recognition of rank for ambulance and other supervisors within the department. This insignia also served as a morale building for the uniformed force.

- Fire Cadet Program - This program provides paid educational and practical experiences to District high school students interested in employment opportunities with the Fire Department. The first class of seven cadets completed training on October 9, 1987.

One never-ending battle Chief Coleman must fight is that of promotions. He has dealt with many hours of depositions and court issues in trying to bring about a racial balance in the Fire Department for the first time. To aid him in this endeavor, he was required by law to submit an affirmative action plan. One might ask what is the reason for affirmative action. Clearly, affirmative action is a tool, which includes any measure, beyond simple termination of discriminatory practice, adopted to correct or compensate for past or present discrimination or to prevent discrimination from recurring in the future.

The union challenged his plan with a lawsuit that has prevented promotions and hiring for the last 4 years. Nonetheless, over this time span Chief Coleman not only promoted more blacks in higher ranks than the department has ever had but, also , non-uniformed women and other minorities. Many staff members believe the uniformed promotions to be the real basis for recent criticisms.

The issue of the Ambulance Service is one that has been prevalent for many years. There were always periodic slow responses and ambulances getting lost. Previously, the service was free to citizens and the chances of being criticized for a death resulting from possible slow response time or location was minimal. It's different with a paid service. No justification can account for loss of life but the problem in itself is still one of the systems, which unfortunately the Fire Chief inherited. There are several points you should know.

A task group comprised of members from the Mayor's Advisory Service Committee requested permission from the Fire Chief to conduct a study of the ambulance service, which was granted in good faith. Consequently, an unofficial draft copy of the study was released to the media in September 1987, which marked the beginning of continuous criticism of our ambulance service. The study neglected to convey that the District's ambulance service responds to approximately 120,000 calls a year on a 24-hour basis (with 21 ambulances), 95 percent of which are handled accurately. These ambulances serve a population of 627,400.

The number of chiefs in the Emergency Ambulance Bureau over the past year also raised some concern. The change in chiefs in no way speaks to the lack of managerial dimensions, but rather to necessary personnel changes made in the department such as promotions, reassignments, etc.

We now have a civilian ambulance director whose reporting authority has been transferred from the Fire Chief to the City Administrator. Frankly, I am appalled at the praise this individual receives for simply carrying out plans of improvement that were originated in the Fire Department prior to his appointment. Each year Chief Coleman requested additional funding for training and equipment; the Fire Department's budget was solely cut. Yet because of the recent cry for better ambulance service, our budget is now increased by $4 million. If adequate funding was made available to Chief Coleman when requested, our ambulance service would already be second to none. Managing the ambulance service requires special training and experience in the D.C. Fire Department, and adequate funding.

Chief Coleman is a man of strong integrity and has voiced his commitment in the ambulance director's reporting authority. Chief Coleman frequently meets with his staff detailing the support functions that the Fire Department will continue to carry out in furthering the Mayor's goal to improve the delivery of ambulance services.

The Washington Post and a select group of local newscasters add tremendously to negative criticism geared towards Chief Coleman. Ninety-nine percent of the articles printed and news aired are erroneous. This creates a false impression of the Fire Department and its leader that can have a significant impact on the public. The correcting of certain misinformation or false data about the Fire Department that appeared in print or was broadcast is part of Chief Coleman's responsibility as an executive. He has prepared rebuttals to news articles, letters to the editor, letters of retraction, all of which are never published. Writers tag a fancy title to an article and print what they want disregarding truth and veracity.

Yes, the Fire Department has problems as any organization. But our number one problem is not in the mechanics of the department. It's the "fashionable" negative criticism of Chief Coleman based on myth, not fact. The Fire Department needs public support and positive, not punitive, approaches to its problems.

Chief Coleman was proud of the fact that the International Association of Fire Chiefs Convention was held here in the District of Columbia, in August of this year. However, it was not without a notice of controversy that was noticeably perplexing, upon seeing the printed brochure for public distribution, with the host chief (Theodore Coleman's), picture not appearing on the cover. This had been a standard procedure for the host picture to appear on the cover since the inception, (1970's), but not this time. Chief Coleman would have been the first black face to be on the cover of the program.

Chief Coleman showed his leadership qualities by not disrupting the planned programs for the convention, as many voices from numerous quarters suggested that he do so. He was more concerned that the citizens of the city would not see their Fire Chief's picture, the one who had done so much for the city, and their care. Also, many Fire Chiefs from all over the world asked who was the fire chief, and wanted to meet him. Equally important was his having to suppress the anger of his immediate family, who felt that he was disrespected intentionally.

Aside from Chief Coleman's regular duties as Fire Chief, he's a member in good standing with the following organizations: International Association of Fire Chiefs, International Association of Black Professional Fire Fighters, National Fire Protection Association, Federal Emergency Management Association, Council of Governments, American Management Association, National Foundation of Black Public Administrators, and the Police Boys' and Girls' Club. He has also received awards from the:

212

- Eastern Branch Kiwanis Club
- United Black Fund Community Service
- Project Harvest
- D.C. Council on Clothing for Kids
- NAACP, Region C Youth Association
- Hotel Security Officers Association
- Neighborhood Medical Association, Inc.

Chief Coleman has a moderate open door and open phone system in his office. He is glad to talk personally to politicians, the public, and firefighters at most any time. And if firefighters in the station can't come to see him at Headquarters, they just meet him at their stations. Chief Coleman likes to monitor radio frequencies when he's at home, and go to bigger fires to take command and watch how the companies carry out their tasks, and to see the department's safety and training measures being used.

Personally, I feel too much down talk has been aired concerning the Fire Chief not to share the other side of the coin. A man who has devoted much of his life to a career in the fire service deserves better.

I applaud Dr. Calvin Rolark who hosted the Sound Off Program on February 4 inviting the citizens to call and state whether the Fire Chief should resign or stay in the Fire Department. The total tally was 100 to 1 to stay. And, again on February 5 several district officials, community groups and leaders, local churches, officers and members of the Fire Department, and more than 100 citizens joined together in front of Fire Department Headquarters in a support rally for the Fire Chief.

I have worked closely with Chief Coleman for 6 years and I can attest to the fact that he is a man that possess strong leadership ability, that stands up to conflict, projects a strong image, communicates with impact, builds personal power, works effectively with both men and women, and one that gets known. Contrary to the picture that the news critic paint, our Fire Chief is number one.

And now the son of Chief Coleman, will speak; Andre' Coleman.

Andre' Coleman

Good evening, I am Andre' Coleman, the oldest of the children of T.R. and Uvaghn Coleman. I would like to present the other members of the family: Teddy (Theodore Coleman the Third), Barbara Perkins, Michael and his wife, Courtney, and my baby sister, Yvette.

Being the oldest, I have the honor of bestowing upon my father how much we love and admire him and his achievements, through his career in the District of Columbia Fire Department. It was instilled in us that you can achieve whatever you have your heart set on. He has shown us with perseverance, determination, and hard work that you can obtain your goal. I can remember every time daddy went to take an examination, how quiet we had to be. I know now that there is a pot of gold at the end of the rainbow. The Bible says, that whatever your heart shall be, that's where your treasure lies. He had a goal, his heart was set on it, and he achieved it. We all know what that goal was, being Fire Chief. I believe that when you have the courage and determination, you can reach your dream.

I personally admire my father for having the courage, and he had a lot of courage and determination to become the Chief of the District of Columbia Fire Department. He lived his dream, and I know as well as you know that he surely will be missed. I would like to read to you a little poem [*If* by Rudyard Kipling] that reflects my father:

> If you can keep your head when all about you
> Are losing theirs and blaming it on you;
> If you can trust yourself when all men doubt you,
> But make allowance for their doubting too:
> If you can wait and not be tired by waiting,
> Or, being lied about, don't deal in lies,
> Or being hated don't give way to hating,
> And yet don't look too good, nor talk too wise;

If you can dream — and not make dreams your master;
    If you can think — and not make thoughts your aim,
    If you can meet with Triumph and Disaster
    And treat those two impostors just the same:.
    If you can bear to hear the truth you've spoken
    Twisted by knaves to make a trap for fools,
Or watch the things you gave your life to, broken,
    And stoop and build'em up with worn-out tools;

    If you can make one heap of all your winnings
    And risk it on one turn of pitch-and-toss,
    And lose, and start again at your beginnings,
    And never breathe a word about your loss:
    If you can force your heart and nerve and sinew
    To serve your turn long after they are gone,
    And so hold on when there is nothing in you
    Except the Will which says to them: "Hold on!"

    If you can talk with crowds and keep your virtue,
    Or walk with Kings — nor lose the common touch,
    If neither foes nor loving friends can hurt you,
    If all men count with you, but none too much:
    If you can fill the unforgiving minute
    With sixty seconds' worth of distance run,
    Yours is the Earth and everything that's in it,
    And which is more; you'll be a Man, my son!

I love you dad!

This next speaker is the brother of Chief Coleman, James Coleman.
He is a minister, and formerly served in the church of the House of
Prayer in Washington, D. C.

James Coleman

To the master of ceremony, distinguished honoree, the head table, and guest, good evening. I am honored that I've been called upon to pay special inspirational tribute to a very fine brother and friend, who is ending a career, and embarking upon a new frontier. For the past 36 years, he has dedicated himself to the saving of human lives, out of his love for humanity.

He came from a strong loving family background, which instilled in him love, not only for himself, but for his fellow man. That love not only; we are brothers' keeper that's sat down in the law of God.

First, he served his country honorably. After he served in the Armed Services, he decided to settle in Washington, D.C. He married his childhood sweetheart, and they had three fine sons and two lovely daughters. They are here tonight. Though the head of his house, he chose a career to fight fires and save lives. Many times his family time was interrupted when he was called away on an emergency to help save lives. But, because he had taught his family the value in an emergency to save a human life, his wonderful wife Uvaghn and his five children remained in support of him at his life's work. He maintained his strong family ties, and became a role model for his offspring. It was not the amount of time he spent with his family, but the quality of time he gave that made the difference.

T.R., we are proud of you tonight. It was not easy for him to achieve the goal that he had set for himself. He knew that the journey of a thousand miles began with a single step. He started his career on the back step of a fire truck, on to the office of Fire Chief. He knew that great men are usually common men such as he; although they are not by the ways of life, but by strength in their struggle to ride above the conditions, setting their sights on a brighter tomorrow. If by chance they run into hard bumps, sometime it simply means they are getting out of a rut. T.R. had his share of knocks and bumps in his daily family life, as well as in the Fire Department. He stood as a pillar of strength for his family, and for his men and women in his department.

He met every challenge. The Scripture tells us in Galatians 6:9, *"Let us not be weary in well doing, for in due season, you shall reap, if you faint not.*" He stood firm in the face of adversity, a man of integrity, a man of courage, a man of love, respect for his fellow man, a man of honor.

T.R., we are proud of you tonight. A time for reaping has come. The time to enjoy the fruits of your labor has come, the time to sit side by side with Uvaghn, a time to enjoy travel, to be a grandparent, to be fishing, to play golf, or whatever you choose to do. As it is said in Psalms, "A man *goeth forth to his work, and to his labor until the evening has come.*"

The evening in the fire department has come. The chapter in this book is closed, and a chapter is beginning. And upon this beginning, we ask for you Chief T.R. Coleman, *"Lord will thou forgive me, to strengthen me, above me to keep me, above me to protect me, before me to direct me, behind me to keep me from straying, around about me to defend me, bless thou our father, forever and ever, Amen.* Proverbs 4:12: *As I goeth, thy way shall be opened up, step by step before thee.* Thank You

Our next speaker is Mr. Joseph Yeldell.

Mr. Joseph Yeldell

Ladies and gentlemen, before I introduce the honoree, who is patiently waiting, I do want to re-emphasize as I have said over and over again, in order to be successful, he had to have an awful lot of help at home. Would you all rise with me and greet Mrs. T.R. Coleman. And now ladies and gentlemen, the man you have heard so much about has the chance now to finally say a few words of his own. You welcomed him before; again, I ask you to welcome him by rising with me and greet our Fire Chief, T.R. Coleman.

Farewell Message

At the conclusion of all the festivities that led up to the date and time of my department from the District of Columbia Fire Department as its

Fire Chief, it was a time of mixed emotions. While I sat and listened to the many accolades and well wishes, I approached the podium, acknowledged the designated persons on the dais, and all other guest assembled. At this time, I began my farewell message:

This is my final opportunity to address you, the members of the Fire Department, in my administration; this is an opportune time to say farewell. I would like to speak first of the accomplishments made in our department which would not have been possible without the involvement and cooperation of each one of you.

Administrative Division

The Administrative Division has developed a system of defining staffing requirements for Fire Department programs. With this system, it defined requirements and had them generally accepted by the District Government. The division has also developed a budget system and an appreciation for long range planning and programming within the budget process. A building maintenance program has been developed with the Department of Corrections whereby rehabilitated ex-offenders are trained in construction skills by working on firestations. This program has the dual benefit of providing job training and at the same time maintaining and repairing firestations.

Fire Prevention Division

1. Developed a comprehensive company inspection program.
2. Participated in Rooming House Task Force with DCRA.
3. Expanded the Technical Review Section for fire alarm systems, sprinkler systems and health care.
4. Obtained additional office space giving an improved work environment.
5. Fire Inspectors are now nationally trained at the National Fire Academy for higher standards of professionalism.

6. Smoke detector regulations have been strengthened and smoke detectors are now available for the hearing impaired.

7. Reduction of fire deaths.

8. Safer buildings in the District through strong code enforcement.

9. Adoption of the new BOCA Fire Prevention Code.

10. Increase of Fire Prevention staff from 52 to 67.

11. Increase in inspections, violation notices, fines issued and abatement of fire code violations.

12. Two fire safe Presidential Inaugurals.

13. Creation of four, two-person Fire Investigation teams.

14. Expanded the Fire Investigation and Arson Units.

15. Established a Fire Inspection Unit at the St. Elizabeth's Hospital Complex.

Apparatus Division

1. Decreased down time on heavy-duty apparatus preventive maintenance (PM).

2. Established and improved the ambulance PM program, up-grading all ambulance units from gasoline to diesel engines for longer service life.

3. Completely changed over from 30-minute Scott SCBA to 1-hour Scott SCBA system with a complete PM program.

4. Purchased second new fireboat to improve the waterfront fire-fighting and rescue services.

5. Designed and fabricated portable carwash apparatus for the Fire Department smoke detector giveaway program.

6. All mechanics have been trained and qualified as diesel mechanics.

7. Planned and coordinated all activities related to the Annual Firefighters Parade.

8. Rehabilitated four fire pumpers to save the District approximately $600,000 (the cost of purchasing four new units).

Under our apparatus purchasing program, in order to maintain our fleet and NFPA Standards, the following have been accomplished:

1. Purchased an average of five new diesel powered pumpers, increasing them from 750 GPM to 1250 GPM.
2. Purchased an average of three aerial ladder trucks every 3years.
3. Purchased one 135' aerial ladder with a second one on order at present time.
4. Replaced the 1971 85' Sutphen Aerial Tower with two Grumman 102' Aerial Ladder Platforms.
5. Obtained second Armored Personnel Carrier, both are currently in service.
6. Assisted the Firefighting Division to place in service a Cave-In-Unit.

The Training Academy has worked on countless other projects, including updating existing training manuals and adopting new ones, officer training classes for senior firefighters, fire inspector training, driver-license upgrading, training the entire department on the new style breathing apparatus, and ensuring that each member received a personal face piece after being fit-tested to determine the proper size.

Instruction in the operation of the "Rough Rider" was provided to units, which will house it, and all firefighting companies were taught the methods of operating as a 4-man engine company.

I am pleased to also share with you several accomplishments that are currently in place, which I personally initiated:

Major Accomplishments

- Fire Safety System - This system is designed for District residents to broaden their prospective on how to stop fires before they start. It is composed of three components: A Smoke Detector Giveaway Program, a Fire Prevention Education Program, and a Home Safety Check Program. This overall system is credited for the low number of 10 fire related fatalities in FY-87 as being the lowest number that the District has had in 50 years.

- Community Action Team - A team that coordinates activities of all Battalion Fire Chiefs for ensuring continuous contact between the Fire Department and Advisory Neighborhood Commissions, Civic Associations, and other ethnic groups to exchange ideas in ascertaining the needs of all parties involved.

- Life Safety System - The purpose of this program is to significantly reduce fire related fatalities and increase emergency care for the elderly and disabled. This system establishes a direct link with Fire Department Communications that eliminates the need for a third party.

- First Command Vehicle - A converted medic unit equipped with telephones and radios for continuous communications between units on the fireground and Communications Division.

- Armored Personnel Carrier - Nicknamed the "Rough Rider," this unit is equipped with fire suppression capabilities for attacking hostile fires when working under gunfire conditions. This is the only armored personnel carrier in the country modified for firefighting use.

- D.C. Firefighters' Parade - An annual event that symbolizes heroic acts of bravery and dedicated services performed by firefighters and which is aimed at making Washington a fire safe community.

Minor Accomplishments

- Established Administrative Task Force Program
- Implementation of Managerial Training Program
- Established Analytical Task Force Program
- Initiated the Uniform Identification Shoulder Patch
- Established the Fire Department Color Scheme
- Purchased Hazardous Materials Vehicles

The Apparatus Division is presently operating a 24-hour per day operation and is awaiting approval to fully establish the work schedule with a minimum amount of overtime.

Firefighting Division

1. Increased home survey inspection
2. Informative inspections and fire prevention inspections
3. New units in service: Hazardous Materials Unit, Command Unit, 4 full service Rescue Squad Units, 135 foot aerial ladder truck, 2 102 foot aerial ladder platforms, and in the process of placing in service, a Cave-In-Unit.
4. Engine companies and rescue squads are now staffed with Emergency Medical Technicians (EMT's) and truck companies will soon have EMT's responding to them.
5. As a safety measure, all firefighters now ride in crew cabs.

Communications Division

The Communications Division is the nerve center for the District of Columbia for fire and emergency ambulance service. The number of incidents handled annually has increased by 68 percent since 1983. In order to deal with such increases, staffing has been expanded in operations, maintenance, and management by approximately 50 percent. Training has been improved and will produce dispatchers well qualified to operate in today's highly technical environment. Stress and ergonomics now receive considerable attention.

Further, we have embarked on a major equipment and facility upgrade. The project, already underway, will be completed in 3 years and is expected to cost approximately $19 million. The result will be state-of-the art communications system, fully capable of meeting demands for fast, reliable dispatching, emergency incident command and control as well as on-line management information.

Training Academy

The Cadet Program has been implemented and cadets have been converted to firefighter positions. A continuing program of EMT and CPR training has been implemented to ensure that every firefighter is versed in these skills.

In order to help assure the physical well being of members of the department, the size and scope of the Safety Office have been increased and a Physical Fitness Program is well under way.

Other programs, which are being implemented, are the High Angle Rescue Team, the Underwater Rescue Team and the Cave-In-Unit.

The training Academy has developed a comprehensive building inspection plan; complete with lesson plans and inspection forms so the Fire Prevention Division with their inspection Program.

- Implemented the Apparatus Replacement Program

- Established the Fire Department Honor Guard

- Initiated the Judicial Affairs Position

- Initiated the Medical Officer Position

- Initiated the Labor Relations Position

- Initiated the Operations Research Position
- Established Mass Casualty Metro Drills
- Initiated Area Fire Chiefs Meetings
- Initiated Military Army Uniform Insignia
- Established Firefighting Division Inspections
- Established the Analytical Monitoring Process
- Directed the purchase of the first 135 foot Aerial Ladder
- Initiated Saturday Fire Suppression Training
- Initiated the Institutional Fire Safety Program
- Initiated Free Citizen CPR Training Program
- Initiated Fire Cadet Program
- Initiated the Smoke Detector Loan Program
- Initiated the Speak Up Program

Thank you for your enthusiastic participation in helping to implement these activities and helping the department accomplish its mission.

An excerpt from that old favorite, "The End of a Perfect Day," states, "Do you turn from your work with a smile and feel that its' all been worthwhile?" These words seem to sum up my feelings. To me, the 36 years I have spent with this wonderful organization have been like a perfect day. I am leaving with a feeling of honor and of deep satisfaction. My heart is full of thanks for all this day has brought me, because it has been both rewarding and worthwhile. And, the day to come will not be as long as the day gone by, but it will be enriched by the memories of my association with men and women like you.

Again, thank you for your loyalty, dedication, and support in moving this department forward. Farewell, and may God bring you good fortune, good health, and a full life of happiness.

T.R. Coleman

In 1981, the D.C. Fire Department formed its Honor Guard to accord appropriate respect to deceased members of the force, both active and retired. The Guard represents the Department when firefighters in other jurisdictions are killed in the line of duty, and they have traveled as far away as New York City.

Depending on the desires of the family, this special group stands guard during the wake, serves as actual or honorary pallbearers, carries the national, state and Fire Department colors and plays "taps." If the military is involved, the Guard's participation is coordinated with the family.

The Guard averages 20 ceremonial functions per year. Aside from paying respect to the deceased, some of the functions in which they participate are: helping to open and close the Special Olympics; the National Olympics; marching in the National Firefighters' and local award parades; and carrying the colors at medal awards and D.C. Fire Department promotional ceremonies.

Lieutenant Samuel E. Moten, organizer and commander, has a roster of 25 members to perform these ceremonial functions.

225

Chapter 26

Reflections

In writing my life story, I have endeavored not to intentionally misrepresent anyone or leave anyone out, nor did I intend to harm or disrespect any one group.

I delayed writing of my experience in life because there were so many people who had given so much of themselves, including the ultimate sacrifice (death), in helping to make these United States of America the great nation it is today. They were heretofore referred to in many other terms or names—Negroes, colored people, blacks, African Americans—but they contributed immeasurably in their forced labor and creative ingenuity in numerous categories, and they commanded the greatest of admiration and respect.

Over my life's experiences and knowledge of the people, I get somewhat of an electrical stimulated surge in my head in attempting to remember the names of those persons whose shoulders I stood on in achieving the successes in my life. To all of them, I am eternally grateful. I must interject here: Many of those persons were not of color, and ranged in status from an average concerned citizen to the president of these United States of America. They exhibited the greatest of intentions and moralist of convictions.

One example of a person not of color, who I consider ones' shoulders I stood on, was Officer Devine, a Lieutenant in the Fire Department. He aided me immensely in the early development of my career, and so on. However, the shoulders of Lieutenant Devine were not exactly the same as others that I referred to in this book, but his input in my career points out how people can show sincerity in their relationships with others, regardless of color. He said to me one day, "Coleman, I notice that you are always reading handbook rules and regulations of the department." I replied, "Yes, I am studying for the upcoming examination for Sergeant."

During my off duty days, I would drive one of the few taxicabs that I owned. Business was good for me, especially when my shift rotation allowed me 4 days off. It also allowed me to continue studying while waiting for passengers at various pickup locations. Also, during those times, I had a musical group called T. Teddy Tunes. We would sing and play the blues in nightclubs. However, I gave this up and devoted much of my available time to studying.

One day Lieutenant Devine and I had a general conversation about life matters. At one point in our discussion, he said, "Coleman may I make a suggestion to you?" I said, "Sure." He said, "I suggest that you quit driving your taxi cab, and concentrate on the examination for Sergeant, because I believe that you will make more money future wise in the Fire Department than you will with the cab. Just a suggestion." I said, "Thank you sir." I gave his suggestion considerable thought. I had a wife and four children dependency. Also, it would seem as though my cabs would develop mechanical problems at times when they could have been most profitable. I took Lieutenant Devine's suggestion, and devoted much of my time to studying for the examination for Sergeant, as well as spending more time with my children. I took the examination, passed, and was placed on the eligibility list for promotion. Subsequently, I was promoted in order, based on an existing vacancy.

After I had attained the rank of Captain, and the required number years for retirement, I submitted my papers through the proper channels for its effect on schedule. Subsequently, I was told by the Fire Chief to be sure to have my retirement papers in his office by the close of business on Friday for his approval, and forwarding to the personnel office. Late Wednesday evening, I received a telephone call from Assistant Fire Chief Devine saying that he had my retirement papers on his desk. I told him that the Fire Chief wanted my papers in his office by close of business on Friday for my signature. He replied, you need to come to my office tomorrow (Thursday), and sign the necessary forms, requesting the effectiveness of your retirement. I reported to his office, and signed the papers. During that meeting, Assistant Fire Chief Devine said to me, "Captain Coleman, may

I suggest that you give what you just signed careful consideration, before I have these papers on the Chief's desk by Friday morning. You have an impeccable record here in the department. As a matter of fact, your unblemished record stands out among many, if not all of your fellow officers. You have the potential of moving up higher in the ranking in the department. However, if you insist on retiring, I will need to have papers hand carried to Chief Lewis's office for signature, and in the personnel office by close of business, and not the other way around. Your retirement will be effective on Sunday a.m." Chief Devine and I continued our discussion, and before I left his office, I told him that I wished to rescind my initial request for retirement, and to remain in the Fire Department. He gave me the necessary forms and I signed them. After signing, Assistant Chief Devine said to me, "Captain Coleman, I think you have made a decision that you won't regret, if you continue performing in your duties as you have."

I am confident that those persons who read my life reflections will be more knowledgeable than I about the many African American pioneers. However, I have acknowledged some in this book, especially those persons who were in a military regimentation, since my life work experiences have been in a uniform of some type. I make this point of reference because it has proven to me that physical and mental discipline can be of great strength when confronted with conditions of diversity, be it a follower or a leader of any group of people.

I am hopeful that my comments have been helpful to the reader, especially members of a Fire Department, and that they will be inspired to want to do their best, whatever their assigned duties or responsibilities are within their vocation, to see the workings of their business or a major cooperation/organization, from all aspects of accountability. Hopefully, they will be inspired to review its failures, imperfections, human relations, civic responsibilities, financial accountability, the power driven source, and the personnel depended upon to make it work. This information may be informative to someone who may not have all the answers in going about the business of everyday life in a workplace.

I would stress to anyone, that with any endeavor undertaken, communication and teamwork are essential to becoming successful in accomplishing a mission. No matter the potential of your business, the worth of your product, the dedication of all involved, the word team is key to your success. Of course, you can be your own individual show.

There is a word called 'leadership.' It should not be taken lightly. At a top management level, one should make every effort to satisfy ones' self that the most qualified persons should be selected to fill all key positions. They should be capable of addressing the greatest of all challenges, and that is to be capable of channeling the awesome power of people. It is so well known that the most serious problems are people problems.

Any and all persons who occupy key positions should be good motivators of people to show interest in their positions. If people are unhappy, unmotivated, or have attitude problems, the primary objective will undoubtedly suffer. People generally do not like to be made to do their given assignments, and there must be some supervision given by more experienced people. No one should be valued less than the material fixtures in the work place.

Too often, management personnel think that success and profitability of the business come first, and the success of the employees are secondary. To me, it is the reverse, showing people how they can succeed helps them achieve personal success. Your success and the success of your department or company can be realized when you realize your most important asset is your people. It is said that managers "manage things, leaders lead people." Strong leadership makes all the difference in the careers of people who work with you. Being a good leader does not come naturally to most people. Sometimes it may be necessary to put the need of others before yours.

Here are some of the basic principles that I found quite helpful while moving up the ladder of added responsibility in the Fire Department. Everybody wants to be somebody. Learn to recognize that desire and determination account for success as much as education and background. By listening and talking to others, I applied this philosophical concept that I practiced:

➢ Treat people with dignity by adding the human factor.

➢ Build positive relationships, mindful of your personal and business life.

- Praise and proper recognition is one of the most powerful forms of motivation. You can be assured that the individuals will want another plaque/certificate to add on the wall.

- Responsibility accompanied by freedom to act independently, and to be judged on their performance, can be of tremendous value.

- Business savvy does not outweigh character and reputation.

- Commitment is the beginning toward greatness. It is the quality that enables one to overcome adversity, and keep moving forward to becoming successful.

- Be adventurous; people who become successful are those who have a cause that they can commit to and believe in.

- Positive attitude. You may not be all things necessary to show absolute success in a business; you should continue moving ahead when things become difficult.

- Establish your expectations. Employees will achieve the standard you set. Expect the best, and encourage standards of excellence.

- Set priorities. Establish what's important. Committing to those priorities is a must.

- Know your destination. If you don't know where you're going, you'll never get there. Clear directions to follow will aid in future success.

- Fear of failing. This holds people back, rather than technical problems. Try learning to eliminate your own fears, and take the approach of becoming successful.

- Paying the price. Determine the price you are willing to pay for success, and realize nothing worth having comes easy.

➤ Be the first to have done it; others will follow, if you set the example.

➤ Develop with quality. To build anything you may as well build it right, quality is important in anything built to last.

➤ Never quit; no one can ever be truly defeated if he never gives up trying. The road to success is filled with disappointments, but the person that makes it to the top is the one who never gave up.

In summation, stay mobile by showing people your physical self periodically, and do not appear distant, or elusive, to not look as though you have no concern for their presence.

You may have noticed that I used the word business a lot in my suggestions. The Fire Department in all aspects is a business, having the primary objective (entwine) of saving lives and property.

Epilogue

I join many Americans every year in commemorating September 11, 2001 (9/11). It was estimated that almost 3,000 people lost their lives in the attack on the Twin Towers in New York City, the Pentagon, Shanksville, Pa., and aboard the hijacked commercial airliners. My heart goes out to and bleeds for the firefighters who lost their lives for trying to do what they were trained to do—saving the lives and property in the communities of which they serve. I would like to think that as a result of 9/11, attitude in reference to race relations brought a new light of love for each other. It has been far too long that race relations have suffered because of negative attitudes being so damaging in our society.

As you know by having read this book, I was in Emergency Operations for 36 years. Many times, I wondered if the department would ever embrace the camaraderie that would stamp out racism, discrimination, and unrest in the D.C. Fire Department. Yes, there have been a lot of improvements made in race relations, but we still have a long way to go.

We seriously need to concentrate more on harmony within ourselves and cultivate our minds on doing things that will ensure the focus on operational safety. Safety is a priority. We must take care of each other. The effort is lost when you are called upon to make life or death decisions. It is my hope that all firefighters remember to place and use all safety equipment because they represent your lifeline. I understand that sometimes when you are called upon to respond that you forget about yourself. That's a natural reaction because of our profession; it has happened to me many times. Once I was caught in a gas explosion. My hands and face were severely burned. I tumbled from the second floor to the first not knowing where I was for a few seconds. But I survived.

My advice to every firefighter is to learn how to protect yourself at all times; and remember always to know where your next step leads. Firefighters have many tools that they employ to protect themselves and each other. The ravages of fire when it is out of control are very dangerous. It must never be taken for granted, even when you are well trained and have the proper tools. I have been there many times. Keeping a cool head in a fire suppression situation is hard to do. But, to do the job and do it safely is the key to what must be done.

Appendix A

A Concise History of the District of Columbia Fire Department

Following is a concise historical review of the District of Columbia Fire Department, as it relates to the beginning of the department, as well as the role of black firefighters. The archives are not clear on the start of either topic. However, the move toward a professional status for the department and for black firefighters occurred during the late 1700s. In the late 1700s, in the area known as Georgetown, the community purchased a hand-pumping engine, and all male citizens were made conscious of their duty to fight fires, at a call meeting in the firehouse.

In the early 1800s, two pumps were purchased for the Treasury Department. During the period of 1803-1804, city officials of Washington approved the names of Fire Directors, one for each of the four wards, which indicates that volunteer fire companies had already been organized and functional. Fire engines of the hand-pump variety were still somewhat crude; however, in 1808, a genuine advancement was made when riveted hose was invented in Philadelphia, Pa. It no longer became necessary for the fire-engine to be brought dangerously close to a burning building. This produced the hose reel, and completely altered the method of fighting fires in large cities.

One night in September of 1812, around 11:00 o'clock, 20 fires were started at one time in different parts of the city. A high wind was blowing, and much of the city would have been destroyed had it not been for citizens responding to fight the fires all night. During the War of 1812, many buildings at different times were damaged by fire, which the fire brigades were helpless to extinguish.

In 1837, Congress granted a charter to the Fireman's Insurance Company of Washington and Georgetown. In 1856, a better organization was created in Washington, by assigning districts to the following companies: Union, Franklin, Northern, Liberties, Perseverance, and Anacostia. It was the duty of that company in whose district the fire occurred to exhibit their lanterns in such a manner that the standing committee could easily find their rallying point.

In September of 1859, a steam fire-engine of the American Fire Company was tested in Washington, D.C. A trial competition was arranged between it and the Franklin hand-engine. The steam engine outperformed the back breaking competition in every phase of the competitive exhibition. At that time, the city fathers were in agreement with the benefits of having a steam fire-engine. In 1860-1861, during the 58th council meeting, a member of that body introduced the first bill for the equipment of a "paid steam Fire Department" as a substitute for the volunteer system.

The turmoil in Washington during the Civil War was such that it demanded a form of organization. On the morning of May 19, 1861, the liquor store next to the Willard Hotel at 14th and Pennsylvania Avenue, N.W., was discovered to be on fire. In the hotel at that time was a General and a number of soldiers called "Ellsworth Zouanes." The soldiers immediately went next door and succeeded as they thought in extinguishing the fire. They were mistaken. After the fire smoldered for several hours, the flames erupted again. This time, the fire spread with fearful rapidity, and the Willard Hotel became filled with dense volumes of smoke. At this juncture, the General dispatched an aid to a Colonel, asking for a detachment of soldiers. It was answered at once by approximately 100 "Ellsworth Zouanes," led by their Colonel. The fire was extinguished successfully.

Until 1864, the Fire Department was wholly volunteer, divided into eight engine companies and two area hook and ladder companies, with a force of about 150 men. There were no order of things, and the board of Alderman, and the Common Council recognized it. The board made the department a paid institution by an act in 1864. Also, during this time, a partially paid Fire Department was organized, consisting of three engine companies (1, 2, and 3) and one metropolitan truck company.

In October of 1864, the fledgling department took a giant step forward. It was at that time the fire alarm system, the forerunner of today's communications system, was founded. It was installed by the Crystal Company. It was called the Cuptal system; and 25 fire-alarm boxes were called "call boxes." Each of these boxes was essentially an automatic telegraph transmitter. It was operated by turning a crank to send code signals to the fire-alarm headquarters. That in turn, identified the location of the box. The headquarters transmitted via telegraph the same number to the firestation. Also, they activated large bells in towers, church steeples which tolled the number of the box, thereby alerting the on-call extra man. The first location of fire headquarters was at 482 Louisiana Avenue, N.W.

The efficiency of the system was plainly demonstrated by a fire in early 1865, near the Riggs bank at 15th and Pennsylvania Avenue, N.W. The alarm was sounded from box 42, at the bank corner, and was telegraphed from the central office to the large bells in 10 seconds. In fact, the bells were said to have commenced ringing before the man had finished giving the alarm from the box. The superintendent, made no secret of the fact that he made only such temporary repairs to the fire-alarm wires as were found to be absolutely necessary to carry through the winter. The reason for this was that the old lines were often carried across house tops, and could not be made reliable by patching up. Each year, he stressed the necessity of speedily rebuilding the entire fire alarm telegraph and adapting the latest technology to improve the system. He pointed out that the latest equipment would enable each company to communicate with the telegraph office or with the office of the Chief engineer by means of signals tapped by hand. He also called attention to the wisdom of making at least three tests daily by company personnel to discover if there were any breaks or irregularities of gongs and alarm bells. Only two fire-alarm operators were on the payroll at this time, so it was necessary for him to work as an operator approximately one sixth of the time in order to make it possible for these two operators to get home for meals, and see their families daily. In 1870, it is recorded that the lone member of the Negro race to be appointed to the department.

The Organic Act of 1871, established the District of Columbia, and provided for a Governor and Legislative Assembly. It also annexed the old city of Georgetown to the newly formed District of Columbia. The Fire Department was officially created on September 23, 1871, and the name has been born by the organization every since. The statute provided that the fire company in Georgetown become the fifth company of the service, and all men would be paid from the city appropriations. This legislative move started the department's growth in the number of companies and in personnel staffing of these units.

During the period of the department's early inception, there was not much of an attraction for young men to come join the service of being a firefighter. The pay was poor, the hours long, and nicety known as "home life" was an impossibility. Working conditions were 24 hours a day, with time only to go home for meals. Duty was six days in a row, with the seventh day off, if no other members of the company were off sick. If so, there was no time off for others. It was a dirty job with much manual labor. The average educated man was not attracted to this type of work. There were horses to be cared for, a task not to the liking of many "city fellows." It is assumed that the typical firefighters of the 1870's possessed a sixth grade education, and on the wage scale classification was lower middle class. There was no prescribed method of appointment. The foreman of each company chose his men and recommended to the Fire Chief those that he wanted to be appointed. Civil Service as an instrument was not to become the process until much later.

In 1872, another Negro man was appointed to Engine 1. This appointment is rather significant, because he was chosen for recommendation by a white foreman.

On September 29, 1875, the fire-alarm headquarters was moved into 486 Louisiana Avenue, N.W., a new building that had been erected two doors from the old location. On this date, the superintendent of alarms, proudly inaugurated the Gamwell System, which consisted of 78 automatic fire-alarm boxes operated on a closed telegraph circuit. This system required at least 25 miles of wiring, and was powered by a battery of approximately 400 cells. In as much as this central fire and police telegraph office handled communications for both services, there were eight police stations designated as fire-alarm stations, which were not connected telegraphically with, nor under the control of the fire-alarm operators. This unhappy arrangement required that when an alarm of fire was reported to a police station headquarters, it was then communicated verbally to another part of the same building by means of a speaking tube, and finally dispatched by the fire-alarm operator over the telegraph system. Under the most favorable circumstances, there were more delays than was desirable before fire apparatus was in route to burning buildings.

In 1881, another Negro man joined the Fire Department. There were no more appointments for a considerable time.

In March 1899, the fire-alarm headquarters was moved from 486 Louisiana Avenue, N.W., to the third floor of the Engine 14, between D and E streets, N.W. Some years later, this company was found arriving before anyone else at box alarms to which they were last due. This was because they were responding on the incoming box, before it was relayed to the companies, then conveyed to the watch desk from headquarters by means of a specially wired circuit that they nicknamed "Willie." Whenever someone entered the firestation who might hear a fire-alarm box coming in before it had been sent out, and who might reveal their secret, members of Engine 14, who wanted to preserve the source of their confidential information, would holler out, "Cut Willie." This was the signal for the watchman to throw a switch that would disconnect the wire (Those boys were characters).

Later that year, the new joker system was put in service, which punched the number of holes in paper tape, corresponding with the fire-alarm box. Prior to the date, an indicator box was used to flash the number of the box received, making it necessary to "clear the indicator" before another fire-alarm box could be received.

In 1902, three more black men (African Americans) were appointed as members of the Fire Department. They were unofficially sworn in. Two additional black men were appointed as members to the department in 1903 and 1906.

In July 1908, the new District Building was formally opened to the public. The chief engineer and the top brass of the Fire Department moved in, as well as Fire-Alarm Headquarters, which was previously located at company No. 14.

In 1919, the department went to two 12-hour shifts each at the platoon system. As result, a great number of men had to be appointed, and several had to be black. Some earlier black firefighters had retired by this time, and the number now stood at 14. It is purported that they asked as a group to be assigned together to a single unit, with black officers to be selected from among them. The strongest motivational factor in their request was the fact that none of them were above the rank of private, and this move would be an appropriate opportunity for advancement of some among themselves to supervisory positions, without regard to the established Civil Service Competitive Promotion System. The Fire Chief at that time agreed to the proposal, and on April 13, 1919, Engine No. 4 was established as the first all black company. The officers selected for positions were not paid for this elevation in rank until January 27, 1921, at which time they were officially promoted to those respective grades. Those three members were the most senior black firefighters, having been appointed to the department in 1902.

As black men were appointed, they were all assigned to the all black Engine No. 4. As the older members retired, others were promoted in their place from within, some being nowhere near the top listing of eligibility on the U.S. Civil Service Competitive Register. However, those men who were promoted were at the top of the Black Register of having the highest marks. The department had a dual promotion system competitive in nature, but separate by color. So then, segregation of the department may be seen as a request to appoint minority individuals. It accomplished an official approbation, perpetuated by appointing new black firefighters to an all black company and promoting black officers out-of-turn to supervise the all black unit.

Black firefighters increased in numbers in the early forties, which brought about a management crisis. During that time, Engine No. 4 had twice the manpower necessary to effectively staff the unit. This was viewed by management as a terrible waste of manpower, so they formed a second black unit as Company 27. In order to achieve this, all the white personnel were transferred out of the station and assigned to other units. The new black company maintained the same promotional practices. The officer ranks were double in all categories.

In 1940, a new building was built to house fire-alarm headquarters. The Superintendent of fire-alarms helped in designing the structure. He had built the first police transmitter. The Chief's cars and fireboat operated on that frequency. World War II caused certain delays in further developments. However, in 1947, trucks were radio-telephone equipped, followed in short order by hose wagons and pumps. The chief engineer at that time decided that the adjunct to the Fire Department should be under its control, and it was upheld. Control of all fire-alarm communications has been vested in the department every since.

By 1949, the number of black firefighters had grown to 62. Two were serving on special assignments. The strength of all Black companies was 30 men, while the White company was 15 men. Another decision was made to create two more all black companies, Engine No. 7 and Truck 10, which were quartered together in the same station. The move was made in November 1949. This time, there was a great difference in the staffing procedure. The Fireman's Association, which later became Firefighter's Association, Local 36, of the International Association of Firemen, AFL-CIO, objected strenuously to the naming of black officers without regard to the Civil Service Competitive System. It was further heed that blacks being promoted on the basis of color was indeed discriminatory. So when the new units were staffed with blacks with the rank of private, the officers were white, taken from the Civil Service Eligible List.

It was a benchmark in 1949 to have blacks and whites operating within the same company, but different to identify as "integration." The Chief in charge of the department could readily see considerable administrational difficulty under the existing system, with more black firefighters coming into the department. He performed a survey among the blacks in the department, and found that they preferred to work in all colored units; however, 4 years later, many black firefighters rejected the validly of that survey.

In 1950, Congress considered legislation to reduce the workweek of the department from 72 hours to 60 hours. The Chief anticipating the enactment, chose Engine Company 13 for the next move, and on July 1, 1950, a number of black privates and white officers were transferred to that location. At that time, there were no black officers assigned to any all white companies. Many groups opposed the making of another segregated company as a step backward. Petitions were sent to the White House, requesting President Harry S. Truman to order the Fire Department to discontinue the practice of segregation on the basis of race. It may seem paradoxical, but the top officials of the Fire Department wanted to put an end to the maintenance of segregated companies based on two factors: the righteousness and morality involved, and the simple economics of trying to operate a department.

In early 1951, the Chief submitted a plan to the commissioners to end segregation in the Fire Department. There was an immediate and an intense objection, and a schism resulted from this proposal. The entity which represented most of the labor force of the department was the Fireman's Association, chartered as Local Union of the International Association of Firemen, AFL-CIO. As an offspring of that parent body, the local's constitution and by-laws contained no covenants nor restrictive clauses prohibiting membership by an individual or class group. The department was not a "union-shop," and members of the department could opt whether to join or not. There were many black firefighters who belonged to the union during the 1930s and 1940s. In the late 1940s, however, several dropped out, stopped paying dues, and became suspended members. As it became clear that the union would stand in opposition to attempt at making additional non-competitive promotions and would aggressively oppose attempts at integration, there were more and more black firefighters who stopped paying dues. They did not see a reason to support an organization that would not fairly represent them or their viewpoint. The stage was now set for labor to achieve solidarity against management over the proposed integration of the firefighting division of the Fire Department.

The Fire Chief appealed to the department to accept his proposals for the termination of segregation. He pointed out the impracticability of maintaining separate companies, as well as the wisdom of accepting full integration. He prepared a full explanation of his position and distributed it to the department. In part, his appeal read, "Many interested persons have come forth to protect my plan of integration, but as of this date, no one has produced a good, just, and logical reason for continued segregation. Most complaints are based on personal prejudices and dislikes, which are not in harmony with the laws that govern our conduct. By what possible stretch of the imagination can one believe that the Chief of the Fire Department holds within his grasp, the power to ignore, waive, or reject the rights of American citizens. These colored firemen are citizens, and their rights under our laws, entitle them to employment without discrimination."

Early in 1952, Local Union countered with a full-blown campaign designed to stop the Fire Chief from achieving his plan. The local appealed to its' membership. To influence members of Congress, it sought the help of the Citizen's Association within the District and conducted

241

surveys of other cities to determine their policies concerning the employ-ment of black firefighters. They mounted an awesome and persistent effort to thwart integration, and they were successful.

President Dwight D. Eisenhower said in his state of the union message in February 1953, "I propose to use whatever authority exists in the Office of the President to end segregation in the District of Columbia." It is not clear, however, to what situation in a broad sense he may have been allud-ing to, when seeing how far the District of Columbia had progressed in the acceptance of integration.

It was a time when the segregation issue was boiling up to a climax. In 1947, a Presidential Committee headed by the President of General Electric Co. called the situation "intolerable." In 1948, the National Committee on Segregation in the Nation's Capital called it a "blot in our nation." In the 1952 elections, both Democratic and Republican parties adopted platform pledges to clear up racism in Washington, D.C. When President Eisenhower took office, the schools were totally segregated; the city had 2,100 white "restaurants and lunch counters" out of 2,200 in the city. Downtown movies were totally segregated; some hotels had 100 percent Negro exclusions, while some would admit Negroes only for banquets or meetings. Recreation areas were segregated. Hospitals had segregated wards; even the D.C. jail was segregated.

In November 1953, a policy order was issued concerning nondiscrimi-nation in the District of Columbia by the Commissioners. Section II stated that each government official shall base all personnel action taken or or-dered by him solely on merit and fitness of the individual without regard to race, religion, color, or national origin.

Startling enough, the same policy order stated that the order "shall not at this time govern the assignment of a colored employee or official to a white company of the Firefighting Division of the Fire Department, or the assignment of a white employee or official to a colored company of the Firefighting Division." It is believed that the reasoning behind this was because of the eating and sleeping arrangements; the general living accommodations during the 24-hour tour of duty would have brought the races in extremely close contact.

While this lack of definitive and corrective action may have pleased some of the rank and file in the department, it was no solace to the heirs, because they still had the almost impossible task of operating under stringent conditions to maintain the company strength of both black and white stations. For example, if a white firefighter retired and a top man on the Civil Service Appointment Certificate was black, the black firefighter had to be sent to an all black unit where he was an "extra" man above their needed strength, and the all white unit from which the man retired, must be "minus" one man from their needed strength, all to maintain separate companies. At that time, there were 1,032 positions in the department, 112 of them filled by black men. Eight of these 112 were officers; one Captain, three Lieutenants, and four Sergeants.

The Fire Chief, who had taken office in 1952, actually issued the order intended to comply with the Commissioner's order by transferring 14 black firefighters from an all black units into previously all white stations on September 18, 1954. This technique was applauded by some, condemned by others, and actually drew another set of battle lines in the issue. The black firefighters did not believe that this was integration, for they were still clustered in all black houses to a great degree; and those black men who had been sent to the all white companies were vulnerable to be discriminated against in many subtle ways. Those 14 men had been sent into seven companies, each company receiving two in a highly visible and evident pattering attempt. Charges of a "quota" system were hurled against management by the black firefighters. The local believed that they had been victimized, and the matter still was not fully resolved.

On February 8, 1955, the representative Democrat from Georgia, and a champion for the Fire Department, introduced a bill in the House of Representatives (HR-3753) designed to set aside the commissioner's integration order and to return the Fire Department to its' previous state of earlier years. Hearings were held on the bill in the District Subcommittee. The Fire Chief testified against the bill, and asked that it not be passed. The bill was not reported out of committee and died in the 84th Congress.

As new men were appointed to the job, they were no longer carried as "extra" men, but were immediately assigned to an all white company so that unit-by-unit, the color barrier was broken. From July 1961 through November 1961, the transition occurred so that by April 1962, there was at long last complete and total integration in the Fire Department, with all houses having a "mixed" staffing pattern.

A full 8 years after the 1954 order to end segregation, it was at last finished. The order had been obeyed in a token manner immediately, but the full spirit had not been fulfilled until years later. Now there were black officers in command of white privates, a heretofore unheard of practice. Despite the fact that it was accomplished by a distributive process of a specific number of firemen and assignments to each station, for all intents and purposes, segregation was dead.

There was something that continued to plague the orderly transition of integration, the return of the black firefighters for acceptance in the labor union. Despite the obvious anti-black attitude of the local during those turbulent years, there were a number of black firefighters who sought to re-establish contact with the local. Those black members who had been suspended for nonpayment of dues applied for restoration to full active membership status. The prevailing constitution and bylaws of the local provided that such applications had to be presented to the Board of Directors for their approval. If an application failed to be approved by the board, it had to be submitted for a majority approval by the membership in a special meeting. In early 1957, several black firefighters applied to the Board of Directors, but were not accepted. The full membership meeting failed to provide a majority affirmative statement, and the applicants were not favorably acted upon. The local officers and representatives of the black firefighters met to reason out the difference of the past. The white membership was adamant against the re-instatement of the black firefighters, and a reasoning approach between them was to no avail.

In October 1960, one black firefighter was accepted for full membership, but it was by no means to be construed as encouragement for more applicants. However, within the next year, there was a greater acceptance of blacks into the local. The white membership began voting to accept members without considering the race factor. The relationship between the union local and the black firefighters remained tranquil for over 6 years.

In late 1968, a group of black firefighters became very displeased with the relationship with the local. Rather than attempt gaining acceptance with the union, they decided to form an association of their own, the Progressive Firefighters Association. In 1968, the number of participants was nearly a 100 or so plus. The union was somewhat dismayed at this action on the part of the black firefighters. Since they were negotiating for an exclusive contract that would afford the opportunity to represent the black firefighters, the local pledged direct support to them as a minority group. This could have been construed as a ploy in luring blacks to swell the membership, in order to strengthen the locals bargaining power, for things that may or may not have been advantageous to minority members.

Black firefighters have the same physical makeup as anyone else, such as eyes and ears. Why they don't use them as an aid in their thinking for the betterment of their condition is somewhat of a mystery to me. They put their head under a cardboard box and pretend they don't know what's going on. Most blacks in this country know exactly what it takes to stamp out racism; i.e., be a loyal citizen, a responsible family man, live by your religious beliefs, stand up to be counted, make your presence known, and don't just sit and do nothing, because that is exactly what is expected of you.

However, having said that, there is a distinction in time of events, and places, that may not fit entirely within my stated comments. For example, there was a fireman that I met, and developed the greatest admiration for during my early years as a fireman. At this time, he had been in the department for some 25 years. He and I talked regularly about issues within the department, of which I learned a great deal from his wisdom/insight. He was serious about the dedication of a fireman's responsibility, even though he had been involved in numerous life saving rescues, that hardly ever any mention was made, in any form. He was a remarkable human being, his name is: Mr. MACIO BARNES. He was truly a professional firefighter.

Upon gaining entrance in the union, I attended a meeting in 1954 or 1955. The union voted to pay the expenses of any white officer accused of discrimination. I immediately submitted the required forms to request the withdrawal of my membership, as many other black firefighters did. There are a number of black firefighters who are still members in Local Union. There are many discriminatory practices used in the District of Columbia Fire Department today against blacks, which gives credence to the theory of a ploy by the local. Many times and often, these practices are allowed to fester, because black firefighters are too laid back, seemingly afraid to make concerns known. It has been said throughout the country, blacks are programmed mentally to accept whatever happens, good or bad, and most of the time, bad. White firefighters in the department cannot be blamed totally for black problems. At no time in the history of the Fire Department have blacks been in a position to aid in the standard conformance of the department personnel.

I reiterate, we can't sit back and blame other people for our problems, we have to do something ourselves. The Progressive Firefighters Association was there as a support vehicle, but was never used to its fullest potential, in spite of the unions non-representation of its' minority members, whose memberships were in good standing.

On December 1, 1966, a radical change was made in the method of assigning units and boxes sounded from the streets. Heretofore, all fire -alarm boxes had definite assignments. These were determined largely by one factor, the nature of the neighborhood. Once that assignment was fixed, that number of companies responded at any hour of the day. Beginning on the above date also, an alarm from a street box, with no accompanying coordination (such as a phone call from the same area), drew an assignment in keeping with the fire-alarm dispatcher's view of this situation. It took into account the time of night, the incident of false alarms at that box, weather, etc. This was the first indication of effect that the rising rate of false alarms had on the department's thinking. These were called "street alarms." The old term "box alarm" was kept, but given a new meaning. When such an alarm was sounded, it now meant the response was chief, four engines, two trucks, and a rescue squad.

A new concept came into being: a "task force" which meant two engine companies, a truck, and a squad, if one had not been previously dispatched. Thus, the department was changing its' old concepts, not so much to shift away from rigidity, as it was a forced accommodation to the increasing number of false alarms.

In 1967, 10 two-way dual channel portable radios were placed in service. In 1968, 31,000 alarms were sounded that resulted in the dispatch and movement of 149,604 times, and the Communications Division administered them all. There were 55,344 ambulance dispatches ending the fiscal year, June 30, 1968.

In 1969, a new radio four-channel base station was installed at Fire-Alarm Headquarters; the station includes one channel on the Mutual Aid Frequency linked with surrounding political jurisdictions. A new radio base station for emergency ambulance service was installed at Fire Alarm

Headquarters. A telephone and amplifier were installed at the Police Department Special Operations Division Headquarters for liaison purposes between the two departments. A 24-hour, eight channel magnetic tape recorder was installed in the operations room, replacing the 30-minute dictabelt machines. A "fire-alarm headquarters security program" was completed. Locks were installed on all gates to the buildings grounds, with a light and emergency telephone being installed at the front gate to the building.

In 1978, communications received its new dimensions telephone, which provided many special features, such as call waiting, call forwarding, and call transfer. The system was interfaced with the Municipal Centrex Telephone System. In May, a communications network linking Fire Department ambulances with all district hospitals was placed in service. This communications system allowed paramedics to communicate from the patient's side with doctors in the hospital emergency room. Voice communications were available to all 15 hospitals. Seven hospitals were equipped to receive telemetry data, that is transmission of bio-medical data. Voice and telemetry communications were transmitted between ambulances and hospital through a communication console at the Fire Department's Communication Center, where a 180-foot tower had been installed to receive radio and telemetry signals.

		Ward	Responses	Transports	Responses	Transports
ADVANCE LIFE SUPPORT (Paramedic) UNITS						
Medic 1*	3420 14th St., N.W.	1	5,736	2,057	6,711	2,313
Medic 3	2425 Irving St., S.E.	3	5,114	2,197	5,941	2,426
Medic 9	1520 C St., S.E.	6	5,629	2,726	6,685	2,662
Medic 11	2225 M St., N.W.	2	5,390	2,167	6,332	2,355
Medic 18*	4801 North Capitol St., N.E.	4	4,879	1,945	5,708	2,283
BASIC LIFE SUPPORT UNITS						
Ambulance 2	1763 Lanier Pl., N.W.	1	6,413	3,728	6,894	3,688
Ambulance 4	4801 North Capitol St., N.E.	4	6,027	3,046	6,027	3,681
Ambulance 5	1300 New Jersey Ave., N.W.	2	6,365	4,035	6,454	4,000
Ambulance 6	450 6th St., S.W.	2	5,205	3,253	5,690	3,493
Ambulance 7	500 F Street, N.W.	2	5,899	3,520	5,925	3,564
Ambulance 8	4300 Wisconsin Ave., N.W.	3	4,747	2,335	4,537	2,247
Ambulance 10	50 49th St., N.E.	7	6,111	3,453	6,188	3,724
Ambulance 12*	2813 Pennsylvania Ave., S.E.	7	6,265	4,237	7,085	4,665
Ambulance 13	2425 Irving St., S.E.	8	6,663	4,347	7,424	4,252
Ambulance 14	1018 13th St., N.W.	2	6,355	3,776	6,870	3,797
Ambulance 15	439 New Jersey Ave., N.W.	2	5,906	3,595	6,307	4,084
Ambulance 16	1520 C St., S.E.	6	6,718	4,623	7,217	4,441
Ambulance 17	1626 North Capitol St., N.W.	5	6,875	4,826	7,457	4,585
Ambulance 19*	2531 Sherman Ave., N.W.	1	6,545	4,466	7,031	4,119
Ambulance 20*	4930 Connecticut Ave., N.W.	3			1,996	1,056
TOTALS			112,842	64,332	124,479	67,435

* Denotes new or relocated units.
Note: Ambulance 20 in service from 4/1/86

FIRE AND MEDICAL ALARMS DISPATCHED BY THE COMMUNICATIONS DIVISION

	Initial Alarm	Additional Alarm	Multiple Alarm	False Alarm	Medical Alarm	Ambulance Alarm	Total Dispatch Activity
October	1,860	45	1	126	2,116	8,296	12,318
November	1,736	55	3	89	1,905	8,204	11,903
December	2,346	50	3	98	2,028	8,234	12,661
January	2,147	41	3	106	1,917	7,953	12,061
February	1,814	32	1	107	1,899	7,332	11,078
March	2,154	58	5	94	2,507	8,202	12,926
April	2,144	42	2	151	2,375	8,391	12,954
May	2,270	48	2	116	2,477	8,844	13,641
June	2,503	45	1	112	2,582	8,869	14,000
July	2,410	32	3	103	2,947	10,575	15,967
August	2,108	31	1	120	2,634	8,925	13,699
September	2,091	39	1	122	2,741	8,832	13,704
Total FY 86	25,583	518	26	1,344	28,128	102,657	156,912

145

248

Appendix B

Seminole Indian Scouts

The Seminole Indian Scouts was one of the United States Army's most effective units during the 1870s, and early 1880s, as they played an instrumental role in taming Southwest Texas. They were men recruited from descendants of Seminole Indian and African American intermarriage. The Seminal-Negro Indian Scouts, as they were called, formed a courageous band of soldiers, enlisted to quell Comanche and Lapin Indian raids along the Texas frontier. They were raised in the Native American culture, and were tough as boot leather. An inscription in the old Fort Clark Guardhouse Museum in Brackettville, Texas, reads, "They did the job they were asked to do, and they did it well."

At its peak, the number of scouts was 42. From 1870-1914, when the scouts disbanded, the scouts never lost a man in battle nor had one seriously wounded. It is a story within itself as to how descendants of the Seminoles arrived in Southwest, Texas. In 1882, after 7 years of bitter fighting against the U.S. Army and Navy in Florida, most Seminoles were relocated to Indian Territory in present day Oklahoma. The tribe took with them many runaway slaves they had protected in Florida. The association of the Seminole slave resembled that of sharecroppers in that they paid an annual tribute of corn to their protectors. Due to the slaves having learned the ways of the white man and their industry and prosperity, they were seen as the real rulers of the Seminole nation. The Seminoles saw themselves as having been transported to a cold climate, naked, without game to hunt, or fields to plant, or huts to cover their poor little children. They were crying like wolves, hungry, cold and destitute.

In 1850, a warrior named "Wild Cat" led a contingent of tribes to Mexico, believing a confederation of Indians would not be bothered. Some of the Afro-Seminoles, as they were referred to, became disenchanted with Mexico and wanted to return to the United States.

After the Civil War ended, the U.S. Government turned its attention to hostile Indian raids in Texas. The 4th Cavalry was headed up by a promising civil war officer. The army also hired Indians to serve as scouts, and promised them land grants if they would do so. The Afro Seminole-Negro Indian enlisted in 1870. Their units' official name was Seminole-Negro Indian Scouts. This was coined of necessity for they had to be classified as Indians to get on the payroll as scouts. In 1866, a congressional act that authorized establishing black regiments included a provision for hiring Indians, but no blacks as scouts. The scouts spoke English-based Creole unique to the Afro-Seminoles (Creole languages develop when two groups with no common language come together and have to find a way to communicate). The scouts were deeply religious. They were not spit and polish military style in appearance; however, their uncanning tracking skills aside, the scouts were ferocious fighters, a fact they proved repeatedly. The scouts were recognized in official reports. They performed very well under their commanding officer. One scout named Payne was jumped by Comanche in Palo Duro Canyon in1874. The officer wrote, "This man has, I believe, more cool daring than any scout I have ever known." Payne was recommended for a medal of honor making him the first African American to earn that distinction. His name unfortunately was misspelled, and officially spelled as Paine. The following year, three other scouts earned their own medals of honor, namely, Factor, Ward, and Payne (Paynes' cousin). The three scouts received honors when riding with their officer in charge, along the Pacos River, where they came upon 25 to 30 Comanches with 75 stolen horses. The foursome quickly tied off their mounts, moved closer and engaged the Indians, "The thing was, the Comanche's had Winchester rifles, repeaters." Soon the battle turned against the scouts, and they decided that an advance in the opposite direction was the best way out. The scouts and their leader dashed for their horses and galloped away. The officer's ill-trained mount bolted from him, leaving him in the Comanche's path. The Seminole scouts raced back for the commanding officer, fighting off attacks on all sides. A Comanche bullet shattered the stock of

Wards carbine; however, the scouts saved the commanding officer. The commander's recommendation resulted in the three scouts receiving the Medal of Honor. Despite the accolades showered upon the scouts, the social norms of the time did not allow Afro-Seminoles to be buried in white cemeteries.

The promise of land grants was never made good either. Thus, when the scouts died, they were interred in a separate plot of land on the fort. The granddaughter and niece of two of the scouts, one an 81-year old schoolteacher, continue to work tirelessly to preserve the Afro-Seminole customs and promote the sense of pride felt by the scouts descendants. The descendants keep the scouts memory alive by coming together when tragedy strikes or to celebrate Emancipation Day in Texas (June 19). Some descendants who toil to preserve the cemetery say, "We weren't slaves, but we sympathize with those who were." "Seminole Day" is held on the third Saturday in September. The descendants of the scouts stroll among the yucca plants and tombstones, remembering and reflecting, giving meaning to the epitaph carved on many of these headstones, "Gone but not forgotten." These brave men gave their all in the most sincere form of dedication as you could ask from any man who professes to be worth his salt as a living human being. However, having said that, they gave so much, and received so little in return.

This is but one more of many immeasurable contributions that African Americans here in these United States have made in the preservation of what is supposed to be a society, with rights and privileges for all of its citizens, even though the country was greatly divided along the lines of race, religion, and above all, a fledging like government that was very much unsettled during the time the Seminole Indian Scouts were formed.

In reviewing the overall circumstance that permeated the governing bodies at that time, the District of Columbia Fire Department was developed in much the same way out of similar conditions. The first African American fireman was appointed in 1870. Another appointed in 1872, and another in 1881. These appointments were similar to that of the Negro Indian Scouts organization. The discriminative makeup of the Fire Department was very much in evidence when I joined in the early 1950s and continued through the 1980s. Fortunately, for me and many other firemen, we were able to rightfully achieve the highest leadership ranking in the Fire Department, based on qualification and longevity. The Seminole Indian Scouts were not as fortunate to be acknowledged as leaders in a specific uniform profile. I cannot over emphasize my respect for those courageous men at times of great diversity. The knowledge about these men gave me much of the inspiration that I needed to persevere in the Fire Department in more difficult times.

Appendix C

No Confidence Theory

Indulge with me if you will in the hypothesis in theory as it applies to a no confidence in the performance of an individual, or individuals collectively. This has to do with a component of the United States military; i.e., the 99th Pursuit Squadron, better referred to as the Tuskegee Airmen, an all black African American combat flying aircraft squadron.

It is almost forgotten now by many Americans that the United States military fought a segregated (troop wise) world war, beginning in 1941, referred to as World War II. This type of application was endorsed by a War College study that was published in 1925, titled, "The Use of Negro Manpower in War." The separation of black and white races has been an official military policy for scores of years. The study was the end product of several years of study by the facility for classes of white students. It concluded that Negroes were far inferior to whites in every way; they lack intelligence and courage, they are morally weak, and of such low character that they should never be mixed with whites. It was a foregone conclusion that they would never be skillful enough to fly aircraft of any type.

Allow me to elucidate of the previous years before World War II in proving the allegations false that was concluded by the War College study. A refutation of this in the category of aviation began in the 1920's and 1930's. A daring black lady learned to fly airplanes in France, becoming the first licensed woman pilot in the United States of America also, the first black male pilot was licensed in America. He also learned to fly airplanes in France. Their accomplishments inspired many black men and women to dream of earning a living in aviation. In 1927, a few more black men began learning to fly, assisted by organized flying clubs. One of the country's first black pilots owned a flight school at Harlem Airport near Chicago, Ill.

The first all black air show was on Labor Day, 1931, in Chicago. In 1933, two highly respected black pilots made long distance flights and round trips from Atlantic City, N.J., to Los Angeles, Calif. In 1938, one of the male pilots became the first black American to fly the airmail during airmail week.

In 1939, a federally funded civilian pilot training program gave black Americans the first military related opportunities to get flight training. By 1941, there were 102 licensed black pilots in the United States of America. However, they continued to suffer the frustrations of hostile receptions of flying fields, and segregated facilities wherever they went. In 1939, an officer named Benjamin O. Davis Jr., graduated from West Point Military Academy. While there, he endured the silent treatment for 4 years by his white fellow students. He was commissioned and went on to serve in the military.

Benjamin O. Davis, Jr., is the son of Benjamin O. Davis, Sr., the first black general to serve in the United States Army. His son, an officer in his own right, was one of 12 aviation cadets to successfully complete the program at Moton Field by training in the BT-13 type aircrafts. Davis was a Captain at the time. However, President Franklin D. Roosevelt issued a directive that blacks had a right to prove their flying skills on an equal basis with whites.

In 1940, the War Department created a black flying unit in the Army Air Corps. Detailed war plans called for the Army Air Corps to activate the all black 99th pursuit squadron, on an experimental basis in 1941. The first black training class reported to Tuskegee, Ala. airfield. The primary training was conducted in Sherman P-17's by black civilian instructors on a contract basis at nearby Moton Field, while basic and advanced flying was conducted by the Air Corps at Tuskegee Army Air Field, by white military instructors. The training was in aircraft type BT-13's and North American AT-6's. Later, a few of the returning black veteran combat pilots of the 99th, and 332nd fighter groups were assigned as flight instructors. There was widespread skepticism that such a program would succeed.

The wife of the president, Mrs. Eleanor Roosevelt, insisted on being taken up in an aircraft flown by a black pilot, an advanced trainee in the 99th Pursuit Squadron. This was viewed by the white officers with great disdain, who did what they could to discourage her. However, she was adamant in proceeding. The flight experience was successful. Soon thereafter, she took a positive unyielding strong stand by insisting to the President, that the airmen of the 99th Pursuit Squadron and their support personnel were truly talented and had the intelligence and skill to operate each and every component of an air unit, from the pilots and all that it took to keep them flying.

In the meantime, Benjamin O. Davis, Jr., had been promoted to Lieutenant Colonel, and assigned to command the 99th Pursuit Squadron. The squadron was at its full strength at the end of July 1942. The squadron was scheduled for several overseas locations. The waiting time was tiresome, and it wasn't until the spring of 1943, that orders were finally issued making it the fourth squadron of an all white fighter group in North Africa, equipped with new P-40's type aircraft, for gunnery and tactical training.

The 99th's first day of combat was June 2, 1943. The squadron was assigned strafing missions against specified enemy targets. Shortly after, bomber escort missions were assigned. The first brush with German fighters came June 9th. Lieutenant Willie Ashley damaged one ME-109 type aircraft in the melee. The first air-to-air victory by a black pilot was scored by Lieutenant Charles Hall against a Facke-Wulf 190 on July 2, 1943. On that day, the 99th also suffered its first air combat loss, two fighter aircraft.

The squadron took part in missions against Sicily, moving there when the island was occupied. In September 1943, the Colonel was suddenly called back to the states to take command of the all black 332nd Fighter Group, which had been activated in 1942, under white officer leadership. Instead of going home with the satisfaction that the 99th had demonstrated that it could perform any job assigned to it, he found an unsatisfactory report had been forwarded to the Pentagon. The report stated that the 99th had demonstrated insufficient air discipline, and had not operated satisfactorily as a team, that its formations had disintegrated under fire, and that its pilots lacked aggressiveness, according to the Colonel. As a result, the 332nd was to be sent to a non-combat area, and the planned activation of the all black 477th Bombardment Group was to be cancelled. Colonel

Davis said bitterly, "In the minds of commanders of the Mediterranean Theater and the AAF, the experiment was over, and blacks had demonstrated their expected inability to perform in combat at the required level of efficiency." The Colonel was furious over this, and went public with his own story of the squadron's achievements. He pleaded his case before the War Department Committee on special troop policies. General George C. Marshall, Army Chief of Staff, decided that an in-depth study would be made of the 99th operations between July 1943 and February 1944, before its fate would be decided. The study, titled "Operations of the 99th Fighter Squadron in the Mediterranean Theater of Operations," concluded "An examination of the record of the 99th Fighter Squadron reveals no significant general differences between this squadron and the balance of the P-40 squadron in the theater." What had no doubt helped solidify this conclusion was that in January 1944, eight enemy fighter aircraft had been downed by the 99th in one day, and four more the next. This assured the AAF allowed the 332nd to continue in existence and prepare for combat.

Colonel Davis says that his opinion was that "Blacks could best overcome racist attitudes in their achievements, even though those achievements had to take place within the hateful environment of segregation; that the Tuskegee Army Airfield should move ahead rapidly, and prove to all to see, especially within the Army Air Corps, that we were a military asset. The war represents a golden opportunity for blacks, one that could not be missed. Our future in the Air Corps would be determined by the account we give of ourselves. It seems as if we have made a number of gains. We owned a fighter squadron, something that would have been unthinkable only a short time ago. It was all ours. The airplane would be the center of the squadron's existence. Furthermore, we will be required to analyze our own problems, and solve them with our own skills; although we might be confronted with problems on the ground by racists who would seek to divert us from our primary mission, I was confident that we could meet all challenges."

The unit moved to Italy in February 1944, and was assigned by the twelfth Air Force to fly cover for convoys, protect harbors, and fly armed reconnaissance missions. In April, the squadron transitioned briefly to Republic (P-47's), then to North America (P51) mustangs, and placed under the 15th Air Force. The all black 99th fighter squadron had continued flying missions, and assigned to the 324th Fighter Group. They were then transferred to Italy, their mustang tails painted a brilliant red for easy identification. The Colonel in command of the red tails, that flew bomber missions between May 1944 and April 1945 that struck oil refineries, factories, airfields, and marshaling yards. They also made strafing attacks on bridges, river traffic, troop concentration, radar facilities, power stations, and similar targets. The Germans called the fearsome pilots of the 332nd "Schwartze Vogelmenschen" or black birdmen. Lieutenant Clarence (Lucky) Lestor, flying his P-51 fighter named "Miss Pelt," made history for the 332nd, when he shot down three enemy fighters on July 18, 1944, while on a bomber escort mission. He said that the danger of the mission had never occurred to him.

The unit received the distinguishing unit citation in March 1945 for extraordinary heroism in action. The group had escorted B-17's during a bombing raid on a tank factory at Berlin and had fought the interceptors that attacked the formation while returning to base in Italy. The citation for the award noted, "Through their superior skill and determination, the group destroyed three enemy aircraft, probably damaged three, among their claims were eight of the highly rated enemy jet propelled aircraft, with no loss sustained by the 332nd group."

Meanwhile, the 99th had been flying dive bombing and strafing missions with the 324th Fighter Group, and scored 17 victories by 1944. They had flown more than 500 missions, 3,200 sorties. In June, the 99th Fighter Squadron joined the 332nd, making the four squadrons, all black group, the largest fighter group in the theater. In the states meanwhile, the four squadrons of the 477th Bombardment Group had been equipped with North America (B-25) medium bombers, but were embroiled in difficulties caused mostly by problems linked to segregation.

On April 5, 1945, the 477th Bombardment Group was transferred from Godman Air Force Base to Freeman Air Base for final training before going overseas. The 616th, 617th, 618th, and 619th were included in the transfer to Freeman's, ostensibly for advanced combat training. They were on the verge of joining those black airmen who had so recently gone before, and were, despite the skepticism of their white counterparts, distinguishing themselves in combat. As fate would have it, the war would end before they joined them. They did fight one battle of greater consequence than any that were likely to have been seen overseas. The white base commander, upon learning that "all colored" units would be attached to Freeman Air Force Base as of April 5, 1945, had previously issued an illegal order concerning the use of the Officer's Club designed to separate "trainees" of the 477th Bombardment Group, all of whom happened to be colored, from the supervisory officers, who were white. As an alternative, the commanding officer at the airfield would offer the colored officers access to the Noncommissioned Officers' Club on the base. The newly arrived officers did not want that action to be taken because, first, it would have displaced the NCO's, and second, the building was referred to as "Uncle Tom's Cabin," and they did not want to be involved in its use. The Negro Officers were very angry. They knew that this denial of their admission to the Officer's Club was arrogant, blatant racism. They knew that their commission gave all of them the right of access to the club, as stated in Army Regulations 210-10, paragraph 19, dated 1940, which opened the Officer's Clubs on all post, bases, and stations to all officers. The regulations clearly stated that being an officer entitled one to membership in any Officer's Club. This was one of the pleasant perks that made military life comfortable. This also gave one access to the swimming pool, dinner specials, bar and locker privileges, dances, weekly out of town shows, and so on. As a practical matter, they were not about to accept accommodations inferior to what the white officers enjoyed. As a matter of principle, they were not going to allow the army to violate clearly stated regulations in an effort to enforce racially motivated segregation of their unit. As a matter of pride, they were not about to endure the disrespect for themselves and what they had accomplished, inherent in this attempt to prevent the free association with the other white officers. They had trained the same, been tested the same, and achieved the same as the white counterparts. And they were prepared to fight and die if necessary to preserve the freedoms to not accept this intolerable violation of their rights and the disregard for their dignity that the order represented. The officers decided to take a stand against segregation at Freeman's Field on the day of their arrival, April 5, 1945. After attending a movie at the base theater, they decided to enter the segregated Officer's Club. Nineteen chose to test this unlawful order by the commander. They lined up by two's, and approached the main entrance. The Assistant Provost Marshall and several military police, which refused them entrance to the club, met them there. They looked at each other in disbelief. During the interim, Second Lieutenant Roger C. Terry, an officer admired and respected, passed by the Assistant Provost Marshall, inadvertently brushing up against him. For that, Lieutenant Terry was arrested, charged, and eventually convicted by general court martial. The others in that particular line were placed under "barracks" arrest. Word spread like wildfire throughout the 477th. Within 2 or 3 days, the figure of those arrested rose to 101 plus 3. The three included Lieutenants Terry, Thompson, and Clinton. Altogether, 104 Negro officers were taken into custody for this act of defiance.

They were cognizant of the fact that theoretically, because this was during wartime, they could be brought up on charges of treason, or mutiny, and subsequently executed. Here was very real danger that this could happen. The idea of people taking aggressive civil rights actions was not yet established in the public mind, much less in the context of the military with its far greater restrictions on personal liberties. Bear in mind that the actions came 10 years before Rosa Parks refused to give her seat to a white man in Montgomery, Ala., and more than 20 years before Martin Luther King, Jr.'s march on Washington. Still, they were committed to the cause, and believed that whatever the outcome, they had no choice but to proceed as they did.

They were released from barracks after about a day and a half, at which time they were asked to read and acknowledge by their signatures that they understood Base Regulation 85-2, which was the illegal memorandum placing the Officer's Club off limits to them. The entire group who had been arrested for entering the club refused to comply with this request and were subsequently rearrested. They continued to be confined to the barracks where they were subjected to extensive and very intense interrogations. Some were interrogated three times. Senior officers from the Adjutant General's Office in Washington, D.C., came to the base for that particular purpose. Their sessions would start around 2:00 a.m. and continue until daybreak. Some officers were transferred from Freeman Field back to Godman on April 28, 1945, still under barrack arrest and still under interrogation by the Senior Officers from Washington. During the interrogation sessions, each protester was called into a small room and questioned relentlessly. Actually, they were trying to find out who their leaders were; they were very anxious to identify them and make examples of them. They tried to break their spirits, but they bonded together. The interrogators urged them to sign a form saying that they would not attempt to reenter the club, and further, to admit that they had disobeyed direct orders. They were told that if they signed the form, all charges would get dropped. They refused and subsequently remained under arrest.

There was a public outcry. The "colored press" was behind them all the way, the Afro-American Newspaper, The Amsterdam News, and mainly the Pittsburgh Courier, just to mention a few. The parents of each officer became involved and eventually all charges were dropped. In August 1955, 50 years later, the Assistant Secretary of the Air Force, announced that the court martial conviction against 2nd Lieutenant Roger C. Terry, who was now president of the Tuskegee Airmen, Inc., had been set aside and that all rights, privileges, and properties have been restored. The 101 plus 3 officers involved in the so-called mutiny would have their letters of reprimand removed upon written request.

Colonel Davis assumed command of the group at Godman Field, Kentucky, in June 1945. The 477th became a composite group until 1947 consisting of the 99th Fighter Squadron and two B-25 squadrons. The 332 Fighter Group was inactivated on October 19, 1945, and the war time saga of the Red Tails ended. Its' record was impressive; more than 15,500 sorties were flown; 111 enemy aircraft, 57 locomotives, and a German navy ship were destroyed. Ninety-five pilots won the Distinguished Flying Cross; and received more than 800 air medals. However, the record of which the pilots of the 332nd are most proud of is, their protection of bombers were successful during its several hundred escort missions, against their enemy fighters.

In 1948, President Harry S. Truman issued Executive Order 9981, which ordered desegregation of the services. The Air Force promptly announced, "It is the policy of the United States Air Force that there shall be equality of treatment and opportunity for all persons in the Air Force without regard to race, color, religion, or national origin." The war record of the precedent breaking 99th Fighter Squadron and the three squadrons of the 332nd's (Red Tails) help bring about the policy.

The policy opened the path for black Americans to serve their country with honor and dignity in peace and war. The nation has benefited immeasurably as a result. In all branches of service, many blacks have attained high ranks to four-star generals and admirals, and the less, both men and women have participated in the NASA space adventures that have had a tremendous impact in many ways on the lives of everything that exist on earth.

Benjamin O. Davis Jr., retired as a Lieutenant General. He resented and opposed segregation. He told a group of elementary students in Arlington, Va., that, "I have felt prejudice all my life up to and right now. Lots of improvement can be made in this nation, and you are the ones who can do it." President Bill Clinton promoted the General to Four Star General.

The story of the Tuskegee Airmen is so profound in the accomplishments of a mission under the duress of the most imaginable. In that, this military organization and the Fire Department are so similar in many aspects. Many employees of the Fire Department were once members in some branch of the United States Military, and were segregated by race. There was a certain method of adaptation into the segregated Fire Department here in the Nation's Capital. This resulted from the military experiences in every aspect; i.e., character stability of a respectable nature, dedication to obligation, respect of uniforms properly worn reflecting patriotism to their country, and their faith in the divine creator of us all, which enable us to triumph over evil and injustices.

With all due respect to this country's governing bodies of today, as well as the major corporate structures leading by positive examples, it can, as Martin Luther King, Jr., said in his speech, "I have a dream, that one day this nation will rise up and live out the true meaning of its' creed: we hold these truths to be self-evident; that all men are created equal."

I would be remiss in not stating the fact that many white men and women of this country, have spoken out, written and championed laws, and acted individually to espouse justice and equality for all.

Civil Rights

Frederick Douglass, the orator, editor, and thinker in the abolitionist cause had lived through the oppressive 1850's when the slaveholders demanded absolute power and went into a war of secession to get it. He had felt the exhilaration of the Civil War as a gateway to liberation, but also the despair of that war in the great cost and suffering in human life. He held the highest hopes for reconstruction, fired by an ambition for national unity and racial equality. Douglass had witnessed one president after another, beginning with Andrew Johnson, followed by Ulysses S. Grant, slowly retreat from the republican principles of Lincoln. However, there were other presidents of this country who among others took actions of retrogression. One that stands out significantly was that of a planned insurrection that white supremacists spent months organizing in Wilmington, N.C., in 1898. A commission concluded that the violence, which resulted in the death of an unknown number of black men, was part of a statewide effort to put white supremacist Democrats in office and stem the political advances of black citizens. The incident is the only known violent overthrow of a government in U.S. history. Afterward, white supremacists in state office passed the laws that disfranchised blacks until the civil rights movement and voting under the leadership of Reverend Martin Luther King, Jr., and millions of supporters of all races, caused the voting rights bill to be an official enactment, and signed by the United States President, Lyndon B. Johnson in 1964. To this very day, the issues of voting and civil rights for African Americans continue.

In spite of behavior of previous presidents, the 34th president of the United States of America, Dwight D. Eisenhower, was a persistent for the voting and civil rights of District of Columbia residents. He and his wife Mamie were forceful in bringing about the desegregation of theaters and eating places here in this city. He ordered the then Attorney General to present a brief in the 1953 Supreme Court case that recognized the validity, and ordered the enforcement of the "lost" anti-discrimination laws enacted by the Legislative Assembly of the District of Columbia during the tenure of a Republican Governor. The favorable ruling thus opened the doors of restaurants, theaters, and other public places to all residents and visitors. He was successful in the passing of legislation for District of Columbia residents to vote for President of the United States, but disappointed that home rule and representation in Congress were not achieved during his presidency. It is noteworthy that his point man in Congress was Senator Prescott Bush, (R-CT), among some others, were valiant in getting the necessary votes for passage of the 23rd amendment granting District of Columbia residents in electoral college. This case was made by republican leaders that residents in the District should have voting representation in Congress, which taxes them, authorize wars in which they fight and die, and subject them to all the laws of the United States of America.

It was undoubtedly a civic duty, as well as a moral responsibility for them to act as they did while in office especially. This was unlike the act in Wilmington, N.C., in 1898, as mentioned earlier; however, it was as memorable as the action of a bold slave who as a defacto pilot, a position no slave could hold, sailed an abducted confederate ship out of a fortified Charleston harbor to a union of ships blockading the Atlantic coast. His name was Robert Smalls. He was made a national hero, and changed the balance of forces in the Charleston area. His action was similar to that of President Eisenhower, Senator Prescott Bush, and others who sought to liberate a group of American citizens from a position of bondage. These were seasoned men in politics as well as life in general, men in powerful persuasive positions, unlike Robert Smalls, 23-year old slave. He disguised himself as the Captain of the confederate steamer war ship, called the Planter, and navigated out of Charleston, S.C., harbor on May 13, 1862, with 16 slaves, 5 women, 8 men, and 3 children. The vessel was considered the best war ship that the confederates had in the harbor with its length of 150 feet, and 46 feet wide, equipped with heavy gun powder. This vessel (war ship) was delivered under a white flag displayed to the union. It was said of Mr. Smalls, a native of Beaufort, S.C., that he knew the location of every reef, shoal, and torpedo in the Charleston harbor and that he could almost navigate the South Carolina coast with his eyes closed.

The actions of Mr. Smalls were skillful, and without a doubt, courageous. His actions in my opinion was not necessarily an act of accomplishment, but a desperate act to free himself, his family, and all others who were on board from the dehumanization that they lived and labored under. Mr. Smalls became a national celebrity having met President Lincoln. He and his crew were awarded more than $400. Additionally, he was awarded $1,500 for having turned the steamship over to the U.S. Navy. He returned to Beaufort, S.C., as a free man in 1865, and started several businesses including a partnership in a railroad.

Mr. Smalls' former slave master lost all of his property during the Civil War. Mr. Smalls bought the house where he had grown up as a slave. When the slave master died in 1875, and his wife's health deteriorated, Mr. Smalls moved her into the house, and into her old room, and cared for her until she died. Mr. Smalls was sent to Washington, D.C., by the Beaufort County voters, as their Congressman in 1875, and served three terms until 1886. He had been successful in helping to get appropriations to improve a small naval fueling station on an island near Beaufort. This small naval facility is now the U.S. Marine Corps Recruiting Training Depot at Parris Island, S.C. Mr. Robert Smalls heroic action during the civil war was so conspicuously in evidence, it was worthy of the highest praise.

The apathy toward Negroes by the union soldiers was enormous. Racial prejudice was deeply ingrained in most whites, North or South. However, southern families and their African American allies were difficult to comprehend, considering the region's history. Most often, a sincere devotion through affection existed between the master and slave relationship, and a divisive wedge of federal reconstruction did not dissolve it. The nostalgia for the former lifestyle of home, and the practice of a traditional Christian way of life, may explain why men and women of color could support the southern confederacy, while anticipating the impact on the immediate emancipation. Although black confederates were not fighting for self enslavement, they sincerely believed that their ultimate freedom, prosperity, and family achievements in the future lay beyond the south side of the Mason Dixon line.

There were other Negroes in the state of South Carolina, who had served in the civil war, while being considered as slaves. Many lost their lives in battle while engaging the federal soldiers. Some lived long lives after the war, maintaining their respect and dignity. For example, consider the obituary notice for funeral service for uncle Charles, 95, Negro slave, and veteran of the Confederate Army, will be held Thursday afternoon at 3 o'clock at the Colored Methodist Church. He drew a Confederate Pension, having fought through the war with his young master who was killed, and later supplied money to take his second master, a Confederate officer, and himself to St. Louis.

One other person that was a confederate soldier was Jerry, a confederate veteran, who died in 1905. An article above the gravesite memorial service held for this veteran read, "Group Honors Ex-slave Who Became Black Rebel, Confederate Heirs See No Irony in Tribute. We're not honoring this man because he is black, remarked one of the memorialist. We are honoring him because he was a Confederate Soldier."

Having mentioned the historic acts by these three men, it is without a doubt that they mirrored the actions of thousands of black men and women who served on the southern side of the War Between the States, specifically, men. There may have been a difference of opinion as to how many served in the military, and whether or not they should be referred to as Confederate Soldiers. The fact is, they were given to wear the same gray uniform as other soldiers, as well as the appropriate weapons, and acted under orders to engage the enemy, such as the blue uniform Yankee Soldier, to kill or be killed. Regardless of an official sanction by the Confederate Congress, they were soldiers, and acted responsibly, as well as courageously. The Congress approved the recognition of all black companies to be viewed as enlisted persons, had that not been done, they may have not qualified as Confederate Soldiers.

Now that an internal situation has been explored to a conclusion; however, it seems not yet over. In 1866, the confederate state of South Carolina had a short-lived pension application for disability compensation, and in 1887, pension relief was granted for regular Confederate Soldiers. It was not until the year 1923 that a legislative act was approved to provide pensions for certain faithful Negroes, who were engaged in the service of the war between the states. Under this legislation, having served loyally as "servants, cooks, or attendants," they were eligible for a pension. However, qualification required at least 6 months of service, and recommendation by the County Board of Pensions. These pensions were not to exceed $25 dollars annually. It was said that too many blacks applied under the 1923 Act. The following year, the act was amended for South Carolina residents who had served for at least 6 months as "body servants or male camp cooks." Pensions did not exist for all counties.

In 1924, the state appropriated $750,000 for white pensions, and expended $744,672.85. During the same year, the state designated $3,000 for black pensions, and $2,840 was expended. Black pension applications contain useful incidental information. Some name changes took place after emancipation. None of their names appear on official master rolls or in military records. Were it not for their pension applications, the employment and confederate service of these black South Carolinians would have been lost forever, and entirely forgotten. Black confederates were a bold and adventurous lot. Racial inequality and social injustice are also a part of southern history, so is historic fidelity and commonality shared by many different races and creeds, in the south. The wealth of information compiled and well documented about the black confederates should not be forgotten.

Again, I have mentioned incidents concerning the confederate/Negro military participation, the notion of these named persons being forever lost. Here is a contrasting situation, depending on one's views. A Fire Chief in Washington, D.C., egregiously took it upon himself to usurp the established Fire Department regulation that established the time period for a Lieutenant to take the exam for the rank of Captain. The time period was 1 year, but he changed this to 2 years. He acted in a way to usurp the law enacted by the City Council. As a result of the Chief's decision, I was successful in encouraging all officers that were affected by this decision to accompany me in filing a complaint. As a result, the City Council reversed the Chief's decision. There were nine Lieutenants eligible to take the exam, and three passed, including myself. Subsequently, we were promoted to the rank of Captain, having been selected from the Fire Chief's most eligible list by rating score. The other eligibles did not rate high enough to be selected. I believe the Fire Department here in Washington, D.C., would not be the highly rated organization that it has become had the promotions through timely testing process been obstructed.

I am fortunate to have attained the rank of Fire Chief, and I am proud to have promoted staff in all subordinate ranks, prior to, and since then, as well as persons of the Administrative staff. One of my most memorable accomplishments was to ensure female Administrative staff receives upward mobility consideration. During my tenure as Fire Chief, I worked tirelessly in an attempt to eradicate all forms of discrimination in the department.

The order given to me by the Mayor to implement an Affirmative Action Plan for the department had some success, in spite of much opposition from various individual groupings within the department. In as much as a plan of action was put in place, the Black Fire Chief's Resource Book, dated 1985, reflected a pattern of structural person positioning in State/Local Fire Departments in which they could possibly revert back to those times, when racial discrimination was the practiced norm in hiring and promotion. A continuous vigilance should be to ensure that any form of discrimination not be tolerated by persons in positions of responsibility.

Promotions

Pictures of Persons Promoted Through the Ranks by Me
from the Eligibility Roster during My Tenure

Donald Edwards

March 3, 2008

Dear Chief:

I must apologize for not responding to your request earlier. I have been occupied
with a project with the company that I work with. However, I would like to
congratulate you on this project and wish you much success. Enclosed is the
photograph that you requested. My dates with the D.C. Fire Department are as
follows:

Date of Appointment: 04-19-69
Date of Appointment as Fire Chief: 07-27-97
Date of Retirement: 11-30-99

It was a great pleasure and rewarding experience to have worked with and for you
during your career. You were indeed an inspiration and a reason for my success. If
I can be of any further assistance please contact me.

Sincerely,

Donald Edwards
Fire Chief (Retired)

Donald Edwards
8/1//1997 – 11/30/1999

269

Rayfield Alfred

12/1988 - 6/1993   D.C. Fire/EMS   Washington, D.C.
Fire Chief
- Advised the Mayor, City Council on all matters related to fire

1/1984 – 12/1985   D.C. Fire/EMS   Washington, D.C.
Battalion Fire Chief
- Public Information Officer (PIO).
- Responded to the scene of various incidents, i.e., White House, U.S. Capitol or residential structures which attracted news organizations, and coordinated the release of information.
- Ensured the department's education/awareness efforts were designed to fulfill the needs of the entire community.

12/1982 – 1/1984   D.C. Fire/EMS   Washington, D.C.
Captain
- Medical Services Officer. Worked with members of the Board of Police and Fire Surgeons in scheduling and monitoring activities of members of the department who were temporarily disabled by injury or illness.

2/1982 – 12/1984   D.C. Fire/EMS   Washington, D.C.
Captain
- Special Assistant to the fire chief.
- Advised the chief on political matters.
- Performed internal investigations.
- Responsible for administration, implementation and coordination of projects assigned by the fire chief.

5/1978 – 1/1982   D.C. Fire/EMS   Washington, D.C.
Lieutenant
- Served with and commanded a fire suppression platoon.
- Responsible for efficient operation, command and discipline in the fire station and at scene of emergencies.
- Temporarily assigned to process employment applications.
- Professor of Cardiopulmonary Resuscitation – University of District of Columbia.

4/1975 – 4/1978   D.C. Fire/EMS   Washington, D.C.
Sergeant
- Instructor at the Department's training academy.
- Prepared lesson plans for training of recruits and veteran firefighters.
- Commanded a fire suppression companies.

9/1963 – 3/1975   D.C. Fire/EMS   Washington, D.C.
Private
- Performed fire prevention and suppression duties.

Everett A. Greene, Sr.

Everett Green
(Deputy Chief)

Theodore R. Coleman
Retired Chief, DCFD
12105 Lihou Court
Ft. Washington, MD 20744

Dear Chief Coleman:

    I was happy to learn of your forthcoming book about your experiences while serving as Chief of the District of Columbia Fire Department. I was appointed December 17, 1961, promoted to Sergeant in 1972; promoted to Lt. in 1974; promoted to Captain 1980; Battalion Fire Chief, 1988 and Deputy Fire Chief in 1991.

    I have enclosed a couple of pictures that I think will assist you with your book. Picture 1 is a picture of me working at my desk as Deputy Fire Chief of the DC Training Academy. Picture 2 is of me at one of the Graduation Program with a Sergeant assisting. Picture 3 is of me standing in front of the Training Academy. Picture 4 is of a few Black Chiefs at a retirement ceremony at Bolling AFB, starting from the left, is Joseph Ford, Terry Francisco, Myself (Everett A. Greene, Sr.), Raymond Thomas, Charlie Carver, Carl Archer, Marston Sloane and Edward Sharpe.

    I hope these pictures will help. If I can help in other way, please give me a call.

           Sincerely,

           Everett A. (Benny) Greene, Sr.
           Retired, DFC, DCFD

Enclosures

T.R it great that you are writing a book to document your outstanding fire service career with the DC Fire Department. There are many African American firefighters who benefitted from your counsel and guidance.

Thank s for being a trailblazer that didn't forget those who would be following in your foot steps.

God speed and good luck with your documentary.

FLOYD A. MADISON
State Fire Administrator

DEPARTMENT OF STATE
Office of Fire Prevention and Control
One Commerce Plaza, 5ᵗʰ Floor, Suite 500
Albany, NY 12231

office: 518.474.6746 · fax: 518.474.3240

email: Floyd.Madison@dos.state.ny.us
www.dos.state.ny.us/fire/firewww.html

State of
New York

Appointed to DCFD.
March 23, 1969.
Promoted to Sergeant.
March 6, 1981-Assigned to Truck 4.
Promoted to Lieutenant.
March 8, 1983-Assigned to Truck 7.
Promoted to Captain*
September 1, 1985-Assigned to Fleet Maintenance Division.
*Allow to take the 1984 Promotional Exam to Captain due to Fire Chief T.R. Coleman changing the time in grade requirement for Captain.
Promoted to Battalion Fire Chief.
June 1, 1991-Assigned to Fleet Maintenance Division.
Transfer to Battalion One #1 Platoon-August 6, 1993
Promoted to Deputy Fire Chief.
March 15, 1995-Assigned to Fleet Maintenance Division
Transfer Firefighting Deputy Chief Office-March 1, 1996
Promoted to Assistant Fire Chief, Operations.
July 30,1997.
Selected to be Fire Chief.
Rochester, New York-June 28, 1999 - September 8, 2007
Appointed by the Governor of New York.
State Fire Administrator-Oct 1, 2008-present.

Floyd A. Madison – 1969
(Deputy Chief of Apparatus Division - Rochester, NY)

Hopefully, I have not delayed your project. Your letter was mailed to my old address in Rochester, New York.

Sincerely,

Floyd A. Madison

Theodore R. Coleman

Dear TR,

I received your letter concerning your documentary. It took a little time to find pictures. I'm not exactly sure what information you need about promotions so I'll include all that I have.

03-01-1964 - Appointed
02-07-1971 - Sergeant
10-29-1972 - Lieutenant
08-14-1977 - Captain
10-04-1982 - Battalion Fire Chief
12-02-1985 - Deputy Fire Chief

I hope this one will be sufficient for your needs.

Sincerely,

Carl E. Archer

Carl E. Archer – 1964

February 29, 2008

Dear Chief Coleman,

I'm honored, that you will include me in your forth-coming documentary. It was a pleasure to have served under you. It was you and with the help of other's within the department who greatly influenced and motivated me to achieved things that I thought I could never have achieved. If I can be of any help in assisting you with your endeavor, do not hesitate to call upon me. And would you please let me know the date when the documentary will be published.

Here are my appointment dates, as well as the picture you requested:

Appointed to department 6-11-72
           "   " Sergeant   10-7-79
           "   " Lieutenant 4-81-82
           "   " Captain   11-30-90
           "   " Batt. Ch.   3-19-95
           "   " Dep. Ch.   1-18-98
           "   " AFC-O   1-30-00

Carlton E. Ford

Sincerely,

Carlton E. Ford

P.S.: If needed my E-mail address

274

DISTRICT of COLUMBIA FIRE DEPARTMENT
WASHINGTON D.C.

THE BEARER
HEREOF **CHARLES CULVER**
BATTALION FIRE CHIEF

IS A MEMBER OF THE DISTRICT OF COLUMBIA FIRE DEPARTMENT,
WHOSE DUTY IS THE PROTECTION OF LIFE AND PROPERTY OF THE
CITIZENS OF THE DISTRICT OF COLUMBIA; ENFORCEMENT OF ALL LAWS
AND REGULATIONS PERTAINING TO FIRE HAZARDS AND PUBLIC SAFETY
FROM FIRE; SUCH OTHER DUTIES AS MAY BE DESIGNATED BY THE
MAYOR AND THE FIRE CHIEF OF THE FIRE DEPARTMENT.

DATE ISSUED:

FIRE CHIEF, D.C. FIRE DEPARTMENT

MAYOR, DISTRICT OF COLUMBIA

I.D. NO.

SIGNATURE

Charles Culver

Staffing Patterns

Statistical information regarding the physical make-up of Fire Department staffing pattern in various states, as listed in a Resource Book, dated September1, 1985

CITY OF HARTFORD

FIRE DEPARTMENT
275 PEARL STREET
HARTFORD, CT 06103
TELEPHONE: 203-525-3123

COUNCIL — MANAGER GOVERNMENT

JOHN B. STEWART, JR., FIRE CHIEF

S E P T E M B E R   1,   1 9 8 5

It is with great pride and honor that we have
again put together a Resource Book, containing the names
and necessary other data of all of the known Black Fire
Chiefs.

No doubt we have left out Chiefs and Chief Officers.
If you are aware of any Chief or Chief Officers who should
be included, please send necessary data to my above address.

Next edition will be published in Spring, 1986.

T H A N K   Y O U

John B. Stewart, Jr.

"SMOKE DETECTORS SAVE LIVES "

A. Name _____

Business Address __201 South Acaica, Compton, CA__ Zip _90220__

Business Phone _____

Optional - Home Address _____

Zip _____ Home Phone _____

B. How long have you been Fire Chief?   Since __12-1-71__

C. Is your department   Paid _X__   Part Paid ____

Volunteer ____

D. Population of City __87,045*__   No. of Employees in Fire Dept. _94_

Number of Blacks in City _60,104_ **No. of Blacks in Dept.   _62_

E. Rank Structure: (Number of personnel in each category)

	Male					Female			
	WM	BM	HM	OM		WF	BF	HF	OF
Chief		1							
Assistant Chief									
Deputy Chief	1								
Battalion Chief	2	2							
Captain	9	9							
Lieutenant									
Dispatchers									
Drivers - Paid __ (Engineers)	.3	9	1	2					
Inspectors - Fire Prevention		2							
Fire Fighters	1	17	1	2					
Paramedics/Firefighters	2	10	1	5					
Other:	2	6	2				3	1	

*As of 1-1-84.

** 1980 Census

A. Name _____

Business Address __City of Bryan P.O. Box 1000_____ Zip _77805_
          Bryan, Texas
Business Phone _____

Optional - Home Address _____

Zip _77802_____   Home Phone _____

B. How long have you been Fire Chief?   Since _1972_____

C. Is your department   Paid _X___   Part paid_____

                          Volunteer_____

D. Population of City _55,000_____   No. of Employees in Fire Dept. _86_

   Number of Blacks in City _17.12%_   No. of Blacks in Dept.   _2_

E. Rank Structure: (Number of personnel in each category)

	MALE				FEMALE			
	WM	BM	HM	OM	WF	BF	HF	OF
Chief		1						
Assistant Chief	2							
Deputy Chief	3							
Battalion Chief								
Captain								
Lieutenant	16							
Dispatchers								
Drivers - Paid	20		1					
Inspectors - Fire Prevention	2							
Fire Fighters	35	1						
Paramedics	3							
Other:					2			

279

A. Name _____

    Business Address __390 Cassell Street, Winston-Salem, N. C.__ Zip __27107__

    Business Phone _____

    Optional - Home Address _____

    Zip _____   Home Phone _____

B. How long have you been Fire Chief?   Since __2 Years, 8 Months (July, 1980)__

C. Is your department   Paid _X_   Part Paid____

        Volunteer____

D. Population of City __131,885__   No. of Employees in Fire Dept. __193__

    Number of Blacks in City __52,968__   No. of Blacks in Dept. __37__

E. Rank Structure: (Number of personnel in each category)

	Male				Female			
	WM	BM	HM	OM	WF	BF	HF	OF
Chief		1						
Assistant Chief	1	1						
Deputy Chief	1					✓		
Battalion Chief	4	2						1
Captain	35	2						
Lieutenant								
Dispatchers								
Drivers - Paid	43	2						
Inspectors - Fire Prevention	4	1				1		
Fire Fighters	64	20			1	2		
Other (Fire Marshal) *See Below		1						

F. Are you interested in keeping in touch via Assocation listing?

    Yes _X_   No____

G. Are you interested in knowing more about IABPFF?

    Yes _X_   No____

Office Clerical					1	2		
Service Maintenance	2				2			
Professional	1				2			

A. Name _____

Business Address _702 Madison, Evanston, Illinois_____ Zip _60202_

Business Phone _____

Optional - Home Address _____

Zip _____ Home Phone _____

B. How long have you been Fire Chief? 2 years Since _12/1/80_____

C. Is your department   Paid _X___   Part paid_____

                               Volunteer_____

D. Population of City _73,706_____   No. of Employees in Fire Dept. 11.5

Number of Blacks in City _15,000_ No. of Blacks in Dept.   _12_

E. Rank Structure: (Number of personnel in each category)

	MALE				FEMALE			
	WM	BM	HM	OM	WF	BF	HF	OF
Chief		1						
Assistant Chief	3							
Deputy Chief	1							
Battalion Chief	4							
Captain	26	3						
Lieutenant								
Dispatchers								
Drivers - Paid								
Inspectors - Fire Prevention	5							
Fire Fighters	53	8	1	2	1			
Paramedics								
Other:					2	.5		

F. Are you interested in keeping in touch via Assocation listings?
   Yes _X___   No_____

G. Are you interested in knowing about IABPFF?
   Yes _X___   No_____

281

A. Name _____

   Business Address _46 Courtland St., S.E., Atlanta,_ Zip _30335_

   Business Phone _____

   Optional - Home Address _____

   Zip _30305_   Home Phone _____

B. How long have you been Fire Chief?   Since _1 year_

C. Is your department   Paid _X_   Part paid _____

                             Volunteer _____

D. Population of City _425,022_   No. of Employees in Fire Dept. _1,017_

   Number of Blacks in City _283,158_ No. of Blacks in Dept. _510_

E. Rank Structure: (Number of personnel in each category)

	MALE				FEMALE			
	WM	BM	HM	OM	WF	BF	HF	OF
Chief		1						
Assistant Chief	1	1						
Deputy Chief	2	2						
Battalion Chief	12	6						
Captain	59	47						
Lieutenant	37	30						
Dispatchers	1	4						
Drivers - Paid	100	99		1	1	1		
Inspectors - Fire Prevention	9	8						
Fire Fighters	179	241		4	6	6		
Paramedics	21	14				1		
Other:	9	26			3	23		

282

A. Name _____

   Business Address ___1605 Grove Street - Oakland___ Zip _94612_

   Business Phone _____

   Optional - Home Address _____

   Zip _____ Home Phone _____

B. How long have you been Fire Chief?   Since __1981__

C. Is your department   Paid _X_   Part Paid ____

                        Volunteer ____

D. Population of City _339,337_ No. of Employees in Fire Dept. _510_

   Number of Blacks in City _156,095_ No. of Blacks in Dept. _117_

E. Rank Structure: (Number of personnel in each category)

	Male					Female			
	WM	BM	HM	OM	NAM	WF	BF	HF	OF
Chief									
Assistant Chief	3								
Deputy Chief	1	1							
Battalion Chief	9	0	2		1				
Captain	41	5		1					
Lieutenant	55	7	1			1			
Chief's Operator	10	4							
Drivers - Paid	72	5	5	1	1	1			
Inspectors - Fire Prevention	4	5					1		
Fire Fighters	156	88	15	6	3	4	1		
Paramedics									
Other:									

A. Name __Theodore R. Coleman_____

   Business Address __1923 Vermont Ave., N.W., Wash., D.C.__ Zip _20001_

   Business Phone __████████████__

   Optional - Home Address _____

   Zip _____   Home Phone _____

B. How long have you been Fire Chief?   Since _March 18, 1982_

C. Is your department   Paid _X_   Part paid _____

                          Volunteer _____

D. Population of City _627,500_   No. of Employees in Fire Dept. _1,535_

   Number of Blacks in City _73.1%_   No. of Blacks in Dept.   _678_
   (Blacks and other minorities)

E. Rank Structure: (Number of personnel in each category)

	MALE				FEMALE			
	WM	BM	HM	OM	WF	BF	HF	OF
Chief		1						
Assistant Chief		1						
Deputy Chief	3	2						
Battalion Chief	24	7						
Captain	55	10						
Lieutenant	78	33						
Dispatchers	21	5	1		1	5		
Drivers - Paid	231	109			1	1		
Inspectors - Fire Prevention	12	28						
Fire Fighters	275	267			2	12	1	1
Paramedics/EMT	49	84	6		4	48	2	
Other: (non-uniformed)	14	28				34		

Sergeant   52 22
Pilot/Engineer (Fireboat)   2   2

*Note: Paid driver column is fire fighting only.

284

A. Name ▓▓▓▓▓▓▓▓▓▓▓▓▓▓▓▓▓ _____

   Business Address _324 Jefferson Street___ _Topeka, KS_ Zip_66607_

   Business Phone ▓▓▓▓▓▓▓▓▓

   Optional - Home Address ▓▓▓▓▓▓▓▓▓▓▓▓▓▓▓▓▓▓

   Zip ▓▓▓▓   Home Phone ▓▓▓▓▓▓▓▓▓

B. How long have you been Fire Chief? 1 yr. Since___9/1/83___
                                      10 mo.

C. Is your department   Paid_ X _   Part paid_____

                         Volunteer_____

D. Population of City _120,269___   No. of Employees in Fire Dept._243_
                        Approx.
   Number of Blacks in City_12,000_   No. of Blacks in Dept.   11

E. Rank Structure: (Number of personnel in each category)

	MALE				FEMALE			
	WM	BM	HM	OM	WF	BF	HF	OF
Chief		1						
Assistant Chief	2							
District Chief	12	1						
Battalion Chief	4	2						
Captain	47	1						
Lieutenant	48							
Dispatchers	4							
Drivers - Paid	47		1					
Inspectors - Fire Prevention	4							
Fire Fighters	39	5	5	4	3	1		
Paramedics (EMT)*	60	3	2	3				
Other:								

* These members have been included in the above figures according to their
  rank and should not be added to the overall total.

285

A. Name ▓▓▓▓▓▓

   Business Address _310 East Fifth St. - Flint, Mi._____ Zip _48502_

   Business Phone ▓▓▓▓▓▓

   Optional - Home Address ▓▓▓▓▓▓▓▓▓ _48503_

   Zip_____   Home Phone_____

B. How long have you been Fire Chief?   Since ____1984____

C. Is your department   Paid _X_   Part paid_____

   Volunteer_____

D. Population of City _160,000_   No. of Employees in Fire Dept. _240_

   Number of Blacks in City _50%_   No. of Blacks in Dept. _69_

E. Rank Structure: (Number of personnel in each category)

	MALE				FEMALE			
	WM	BM	HM	OM	WF	BF	HF	OF
Chief		1						
Assistant Chief	1							
Deputy Chief - Fire Marshall	1							
Battalion Chief	6							
Captain	8							
Lieutenant	20							
Dispatchers	2				5	4		
Drivers - Paid	35	4						
Inspectors - Fire Prevention	5	1						
Fire Fighters/EMT	46	54	6	2	5	3		
Paramedics	7	5	1			1		
Other: Arson Inv. Tng. Officer / Asst. Tng. Officer	3							
Sargeants	12	1		1				

286

A. Name_____

   Business Address___4400 Memorial Drive Complex_____ Zip_30032_
                  Decatur, GA

   Business Phone_____

   Optional - Home Address_____ Atlanta, GA_____

   Zip____30311_____ Home Phone_____

B. How long have you been Fire Chief?   Since_July, 1985_____

C. Is your department   Paid_X___   Part paid_____

                     Volunteer_____

D. Population of City_500,000+_____   No. of Employees in Fire Dept._475_

   Number of Blacks in City__23%__   No. of Blacks in Dept.   51

E. Rank Structure: (Number of personnel in each category)

	MALE				FEMALE			
	WM	BM	HM	OM	WF	BF	HF	OF
Chief		1						
Assistant Chief	2							
Deputy Chief	3							
Battalion Chief	11							
Captain	70	1	1					
Lieutenant	57							
Dispatchers								
Drivers - Paid	88	4						
Inspectors - Fire Prevention	3	3						
Fire Fighters	91	48			2			
Paramedics								
Other:	1				2	2		

A. Name ▓▓▓▓▓▓▓▓

Business Address   301 Second Avenue South
Seattle, WA   Zip 98104

Business Phone ▓▓▓▓▓▓▓

Optional - Home Address ▓▓▓▓▓▓▓▓

Zip ▓▓▓   Home Phone ▓▓▓▓▓▓

B. How long have you been Fire Chief?   Since July 15, 1985

C. Is your department   Paid X   Part paid_____

             Volunteer_____

D. Population of City 500,000   No. of Employees in Fire Dept. 1,000

Number of Blacks in City 50,000   No. of Blacks in Dept. 109

E. Rank Structure: (Number of personnel in each category)

	MALE				FEMALE			
	WM	BM	HM	OM	WF	BF	HF	OF
Chief		1						
Assistant Chief	1							
Deputy Chief	7							
Battalion Chief	21	2	1					
Captain	45	7	1					
Lieutenant	126	7		12	3			
Dispatchers	15	1						
Drivers - Paid								
Inspectors - Fire Prevention	15	1			1			
Fire Fighters	554	85	19	35	32	2		
Paramedics	53	1	1	4	1			
Other:								

Fire Chiefs

Shown on the following pages are the official Fire Chief's symbol of Washington, D.C., and photographs of Fire Chiefs since 1871.

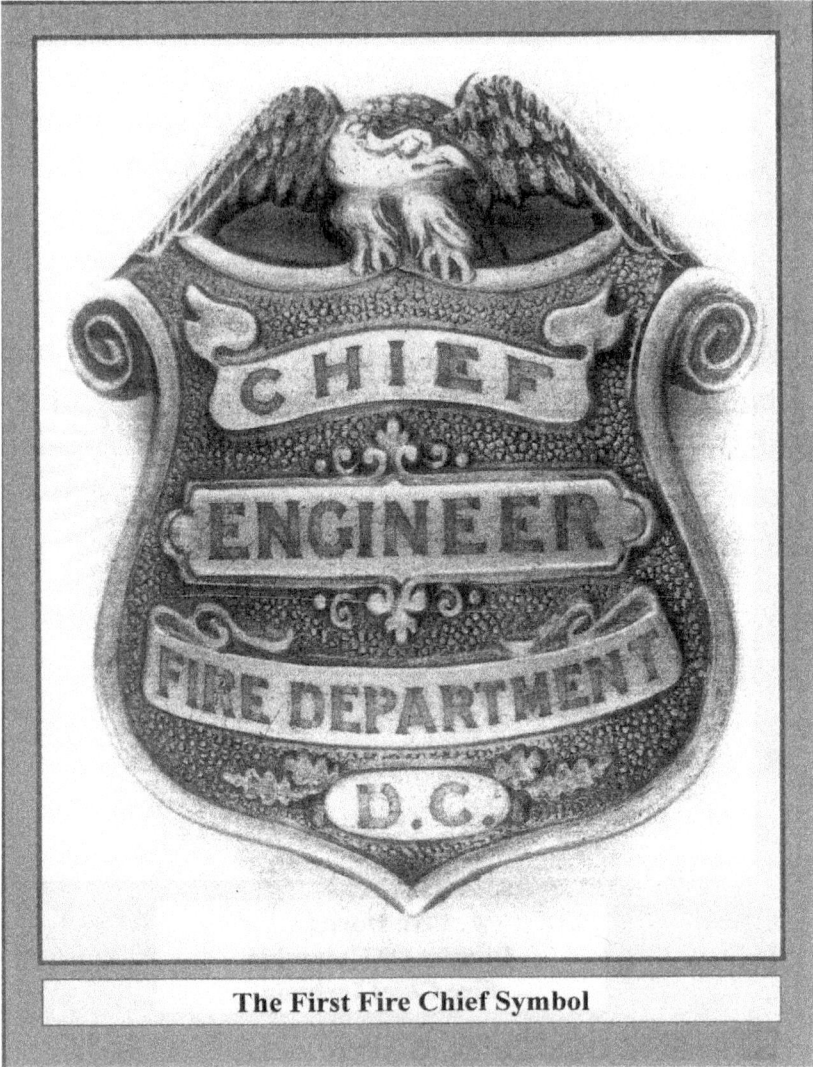

The First Fire Chief Symbol

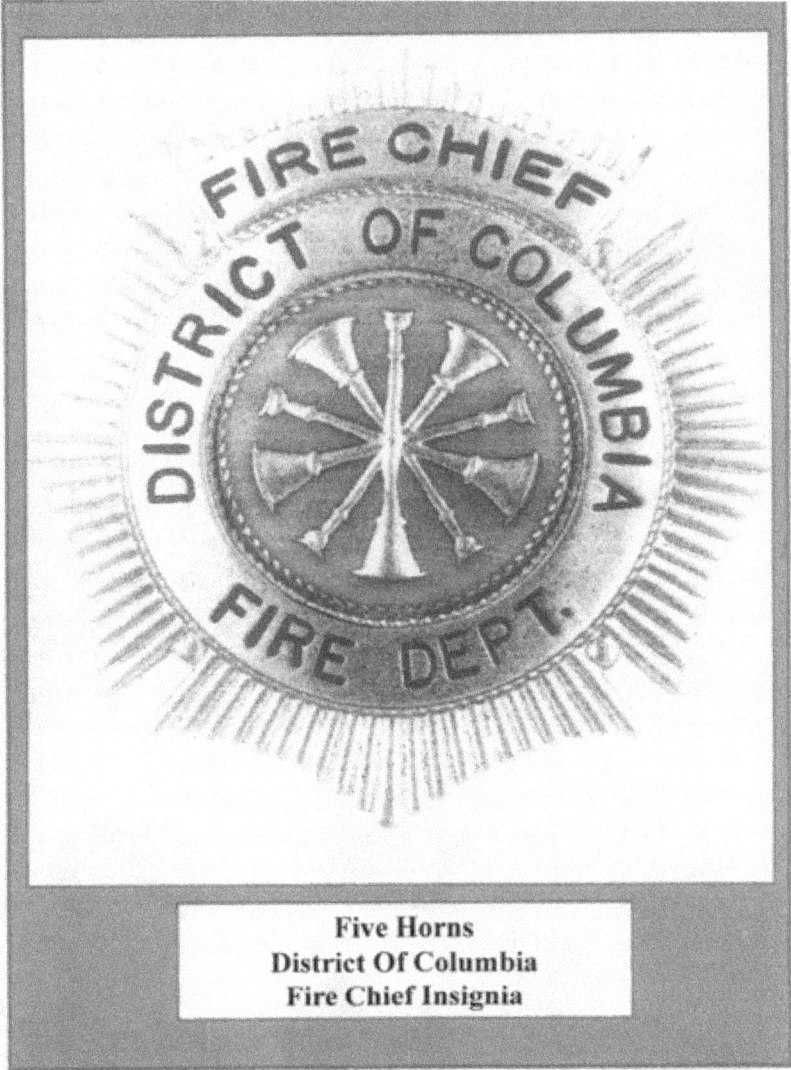

Five Horns
District Of Columbia
Fire Chief Insignia

292

About the Author

As a child, I grew up in the household of my parents, Theodore R. Coleman, Sr., and Essie Mae New Coleman, along with my brother James G. Coleman and my sister Hazel Coleman, in Danville, Va. Eventually I went to my grandparents' home in Ellington, S.C. It left such an indelible image on my immediate presence; in love and affection, which has been a form of sustenance encouragement in my developing life. In our household, I was the oldest son, and was expected to set an example indoors and outdoors, when we played together.

In my physical developing stature, and the ability to take on more responsibility, my father would give me explicit instructions on completing my chores on our farm during the weekends. During the week, I had to study my school assignments and do homework.

Our family house was referred to as the <u>BIG HOUSE</u>. During the time when race discrimination was a common practice, African American people were welcome to stay over during their travels. My father was a deacon in the Baptist faith; he was also a "WM" (Worshipful Master) in the Masonic Order. During some evenings, we would attend church, and sometimes we would attend the church of our grandfather, a member of the Pentecostal faith. Bishop (Sweet Daddy) Grace preached regularly in dress, grace, and style, when delivering his sermon. My brother, James, and I were eight and ten years old, when we would attend the Pentecostal church service with our grandfather. I would sit in the big chair on the pulpit, and attempt to emulate the style of Daddy Grace, and my brother James would sing and pretend to conduct the service, before the congregation arrived. James developed affection for the Pentecostal religion, and through his years of study, he is referred to as Apostle Dr. James G. Coleman, in the state of Maryland.

In my early teens, my work ethic was becoming a part of me fundamentally. So, after I graduated from high school, I went to work in an industrial plant where straw baskets were being made. I became very proficient in my allotted assignment, and was permitted to leave the plant early for the day.

During my time at the plant, I came in contact with people that were opinionated about subject matters during break time, or lunch period. I began to learn early on, that it was necessary for me to give considerable thought before engaging in subject matters that I was not familiar with, and not talk just to be in a conversation.

I returned home after working at the plant one day, and received an official letter, directing me to report to the U.S. Army Induction Center, for military service. I remember what a man said to my father one day, "When that boy get out of school, he won't have to serve in the war, if he can get a deferral approval from the Military Draft Board, based on his services needed on the farm here at home." My father did not encourage me to consider this, so I reported as directed. This was where my parental upbringing began to really be appreciated, with respect for authority in a training process. I already knew the basic principles; because of this, I was able to adapt to the use of equipment to defend others and myself. After receiving training in various facilities for several months, I was marched onto a troop transporting boat, and off to a foreign land (Japan), where I saw the likes of people as never before, however; human. Their basic needs were like any other; however, their culture difference was unique of their own. Upon returning to the United States, I became employed in various businesses, in the Washington, D.C., area, including the Navy Yard.

After marrying my girlfriend, and starting a family a few years later, I needed extra financial income, so I formed a musical group referred to as "T. Teddy Tunes." I was the lead singer. We performed at many clubs in and around the metropolitan area. In the meantime, I took various educational courses, including watch repairing. In addition, I acquired a number of taxicabs, as my family needs increased. My wife had a full time job caring for our children at home.

While employed at the Navy Yard, I applied for acceptance in the D.C. Fire Department, and was eventually accepted. From this point on, I began as a rookie fireman, was promoted up through all the rankings, and finally was appointed to the position of Fire Chief. I held this position for six years, longer than any other D.C. Fire Chief did, in the history of the department.

During my 36 years in the department, I was instrumental in addressing many egregious conditions that were evident in the daily functioning within the department. However; during those years, I am proud of the many accomplishments that were made, and not without the support of so many others, including public officials, and city residents. I believe that the citizens were as proud as I was, at the time.

In my leaving the Fire Department, a void in positive leadership was not in existence, and that included all levels of leadership/staff. As a result, I envision future accomplishments will be even greater than the past.

Please review my farewell message during my retirement ceremony, wherein I detailed many accomplishments during my service as Fire Chief in particular.

Contact Information

To inquire about Fire Chief (Ret.) Theodore Coleman doing a book signing or book discussion or speaking at your event, please contact:

Theodore R. Coleman
12105 Lihou Court
Fort Washington MD. 20744

(301) 292-8875

You may also contact the author by sending an email to le4fo1@verizon.net or by calling (301) 292-5529.

www.ingramcontent.com/pod-product-compliance
Lightning Source LLC
Chambersburg PA
CBHW072055020426
42334CB00017B/1522